Data Governance and Compliance

"Rupa Mahanti has written an essential and ambitious compendium on data governance. She provides a sweeping history on how we got to our present state and grounds her research and recommendations in interviews with business thinkers on data governance. Along the way, she sorts out some of the confusions that bedevil us–corporate governance versus management, IT governance versus data governance, regulatory compliance versus quality–as we try to make smart use of data to meet business goals. Organizations grappling with the challenges of compliance with regulations and standards will find robust ideas to aid their efforts. This book is a great resource for both academic and business practitioners."

—Eileen Forrestor, *CEO of Forrester Leadership Group, Author of CMMI for Services, Guidelines for Superior Service*

"We are all affected by governance in its many varieties—overarching corporate, data, IT and so forth. And yet most of us usually sees only a slice in our day to day work. This treatise on governance—with a focus on data and its enabling role on all other types—is both informative and actionable, ensuring that your governance efforts are grounded in proper intent and designed for both compliance and performance."

—Jeannine Siviy, *Director, Healthcare Solutions, SDLC Partners, Co-author of CMMI and Six Sigma*

"Whether you are a Business Owner, Data Controller, Data Steward, Data Processor, Quality Controller, Cyber/InfoSec specialist, I would highly recommend that you add this book to your reading library to gain a better understanding of the role Data Governance plays in compliance and supporting the success of your business."

—James Seaman, *Director, IS Centurion Consulting Ltd. Author of PCI DSS*

Rupa Mahanti

Data Governance and Compliance

Evolving to Our Current High Stakes Environment

Springer

Rupa Mahanti
Strathfield, NSW, Australia

ISBN 978-981-33-6876-7 ISBN 978-981-33-6877-4 (eBook)
https://doi.org/10.1007/978-981-33-6877-4

This Springer imprint is published by the registered company Springer Nature Singapore Pte Ltd.
The registered company address is: 152 Beach Road, #21-01/04 Gateway East, Singapore 189721, Singapore

Data Governance: The Way Forward

This mini-series of three volumes looks at different aspects of data governance. The volumes do not assume any prior or specialist knowledge in data governance. Data Governance is an important component of the data management discipline, and these books should appeal aspiring information management candidates, aspiring/current BI and reporting analysts, project managers heading information management projects, business and IT consultants, especially those involved with data, university students, professionals in organisations from all industries and sectors who wish to gain a comprehensive understanding of data governance.

To my Mom and Papa, for their unwavering support, dedication, love and encouragement.

To my teachers and mentors, for their guidance and patience.

Foreword by Dan Power

This book starts with a thorough review of the word "governance"—corporate governance, financial governance, HR governance, risk governance, operational governance, security governance, IT governance, and data governance. This detailed discussion of the worldwide history of governance hints at how much ground still remains to be covered in data governance. Areas like financial and IT governance are much more evolved than data governance, a relative latecomer.

The evolution of data governance doesn't take place in a vacuum. It takes place in an era where we're hurtling ever faster toward a data-driven world, where people's "data trails" begin at birth and follow them—for better or worse—throughout their life, where large corporations increasingly compete on the quantity and quality of their data and their ability to make sense of and act on it, where data security has become ever more crucial, and where the world's economy is inextricably linked to the world's data.

Initially, data was treated as the by-product or "exhaust" of business operations, with minimal value in itself. But as technology and business continue to evolve, data's strategic value has become obvious, to corporations, government entities, other organizations, and individuals.

The storage, quality, governance, and management of data is no longer considered solely an IT function. Once the business side of the organization started to realize the value of data, conflicts started to arise between different functional areas, between the business and IT organizations, and between the business and society as a whole. This led to the emergence of data governance as a discipline on the border between business and IT, and as an organized way to resolve those conflicts.

Regulations such as Sarbanes-Oxley, Dodd-Frank, Basel II, and the General Data Protection Regulation (GDPR) have challenged organizations to improve their data quality, and to create controls and formal accountability for data. This requires creating an effective data governance capability, as a collaborative, global, organization-wide function including both business and IT.

As the practice of data governance has matured, there is more consensus on how to define and practice it. Groups such as DAMA International and the Enterprise Data Management Council have defined widely accepted best practices and organizational maturity models. And many organizations have found it critical to align the data strategy developed by their data governance function with the overall corporate strategy developed by senior management.

Since IT governance and data governance are both subsets of corporate governance, it's interesting to see how often they are confused with each other. Of course, both disciplines are focused on improving the company's assets to generate better business value. But, IT governance manages technology assets, and is application- and infrastructure-centric, while data governance manages data assets and is data-centric.

There's certainly overlap between the two. Many organizations have been successful in data governance only when they've achieved a new level of business/IT alignment while ensuring that the data governance process remains firmly rooted in the business, and that the technical aspects of data management are strongly supported by the IT organization.

Data governance is about "doing the right things"and data management is about "doing things right." The dividing between governance and management will vary from one organization to another, and even within the same organization will change over time. This is not a hard and fast line—there are overlaps, and the data governance function (usually in the business) will have to work closely with the data management function (usually in IT).

A new theme has emerged in data governance over the past few years—the difference between "offensive" and "defensive" data governance. Defensive data governance is focused on improving regulatory and contractual compliance, reducing residual risk, increasing data security and privacy, and supporting internal requirements and reporting standards. Offensive data governance is focused on delivering business value, supporting the overall corporate strategy, increasing revenue, satisfying customers, etc.

Some aspects of data governance—improving data quality, tracing data flows (lineage), enhancing data privacy and security, creating a culture of transparency and accountability—benefit both the offensive and defensive sides of data governance.

This book looks at many aspects of data governance, firmly in its context as a subset of corporate governance, and explores what it takes to be successful with the organization, practice, and discipline of data governance, as well as its role in improving regulatory compliance, data quality, data security and privacy, and data architecture.

It provides a great overview of the history of governance around the world, the evolution and basic concepts of data governance, and its large role in driving increased regulatory compliance. For industries that are already highly regulated such as financial services, health care and life sciences, data governance has had a large part in their regulatory compliance efforts for some time now. But other industries will see an increased interest in their data governance and data management practices on the part of regulators. This book provides a great road map for anyone who's just been told by their immediate manager that they are now in charge of data governance for the company, and for anyone interested in the deep and fascinating world of governing today's large corporations, government entities, and non-profit organizations, and their impact on all of our lives.

Hingham, USA Dan Power

Preface

*With proper **governance**, life will improve for all.*

—Benigno Aquino III

I have been working with data for a long time. What has changed concerning data, gradually over time, is the volume and varieties of data, and the tools and technologies which handle data. While the concept of data quality was evident to me and was a topic of discussion starting with when I used data for testing applications, data governance actively came into the picture much later, when I started working in the industry.

Thinking about governance as a child, I used to associate governance with government—authority and control. This was understandable, given that a new government came into power in my country every 5 years or less.

Later, I would also think of governance in association with organizations. Corporate governance has evolved, with a number of factors like takeovers, mergers, acquisitions, stock market crashes, accounting scandals, and globalization, further triggering the need for stricter corporate governance and creation of regulations to improve the accuracy and credibility of corporate information with an intent to prevent financial disasters. Compliance with regulations requires data, and the early drivers of data governance originated as mandates in the compliance, legal, risk, and audit departments within organizations.

While data and governance have been in existence since ancient times, and corporate governance for a few centuries since the industrial revolution, the concept of data governance is relatively new. It has come into prominence in the last few decades. While data has always been important, it has never been captured, shared, or stored in large quantities to require effective organization, processes, policies, accountabilities, and control, until the digital age.

Before the advent of computers and technology, important information was recorded in paper files that were accessible to a limited few and were protected by keeping the files in locked cabinets. With the advent of computers and technology, since processing and storage were expensive, comparatively less data was collected, and it was initially treated as a by-product of an organization's IT systems and application.

Advancements in technologies and processing coupled with cheap storage have enabled the capture, processing, and storage of large amounts of data. Data is no longer a by-product but a commodity and asset that needs to be managed and governed effectively. Data governance enables you to effectively and proactively manage data assets throughout the enterprise, by providing guidance in the form of policies, standards, processes, and rules, as well as defining roles and responsibilities to determine who will do what with respect to data.

Effective data governance is essential for attaining high-quality data. Data quality and data governance have a symbiotic relationship. When I was writing my first book on data quality, I realized I had a lot to convey to readers about data governance that needed a book in its own right. What started as one book is now a data governance trilogy:

- **Data Governance and Compliance**. This first book in the trilogy sets the stage in terms of the evolution of corporate governance, laws, regulations, other forms of governance, and how data governance interacts with other corporate governance subdisciplines. This book goes on to explain how data governance helps in achieving compliance.
- **Data Governance and Data Management**. The second book provides the big picture of data and data governance, including data as an asset, different technical aspects of data, data governance drivers, and benefits. It also covers interactions with various data management disciplines and initiatives, and data governance technology and tools.
- **Data Governance Success**. The third book discusses the different perceptions that individuals and organizations have about data governance, the challenges in implementing data governance, and the key factors that should be considered to ensure data governance success.

These books share the combined knowledge related to data and data governance that I have gained over years of working in different industry and research programs, and projects associated with data, processes, and technologies. I have interacted with professionals spanning the globe and read many books and articles, most of which appear in the references. These books will be highly beneficial for IT students, academicians, information management and business professionals, and researchers to enhance their knowledge and get guidance on their own specific data governance projects.

This series is written primarily for information management professionals, risk management and compliance professionals, data quality practitioners, information management researchers, and students looking forward to a career in data management or governance. Also, this series will be useful for aspiring information management candidates, Business Intelligence (BI) and reporting analysts, project managers heading information management projects, business and IT consultants, university students, and professionals in organizations from all industry sectors who want to gain an understanding of data governance.

Data governance is often overlooked, and its value is grossly underestimated. A lot of people are highly skeptical about data governance. To retain the interests of such an audience, I have conducted interviews with 11 individuals (thought leaders, researchers, and consultants), and their interview responses appear in the last chapter of this book with an intent to share their challenges and experiences with data governance.

Each book has a slight overlap in terms of people, processes, and metrics-related aspects of data governance, though it is the third book that covers these aspects in detail. The technical aspect is a major component of the second book and will be moderately discussed in this book in relation to compliance and minimally discussed in the third book.

Whenever I read a book on a particular subject, from my student days to this day, I find that a book containing a balance of concepts and examples, and illustrations is easier to understand and relate to, and I have tried to do the same while writing this series.

Compliance is one of the biggest business drivers of data governance. As data continues to evolve as well as grow exponentially, and criminals find new ways to break the law, new regulations appear, and existing rules are refined to become more stringent to minimize fraud and risks. The cost of compliance is high, but the cost of non-compliance is higher. Non-compliance costs are those that result when a company fails to comply with rules, regulations, policies, contracts, and other legal obligations.

This book contains seven chapters. Chapter 1 introduces the readers to the concept of governance and corporate governance, the different corporate governance disciplines, and the difference between governance and management. Chapter 2 discusses the evolution of governance and corporate governance. Chapter 3 discusses the subset categories of corporate governance including data governance, the lack of standardization of the data governance definition, the difference between IT governance and data governance, and how data governance ties to other governance disciplines. Chapter 4 discusses a few key regulations. Chapter 5 discusses the impact of regulations on data, the cost of non-compliance with some real-life examples to drive the message home, the role of data governance in being compliant with regulations, and regulatory compliance and data management. Chapter 6 presents concluding thoughts and the way forward. Chapter 7 presents the interview responses.

In case you have any questions or want to share your feedback about the book, please reach me at rupa.mahanti0@gmail.com or https://www.linkedin.com/in/rupa-mahanti-62627915.

Strathfield, Australia
September 2020

Rupa Mahanti

Acknowledgments

Writing this book was an enriching experience and has given me great pleasure and satisfaction, but has been more time-consuming and challenging than I thought. I owe a debt of gratitude to many people who have directly or indirectly helped me on my data governance journey.

I am extremely grateful to the many leaders in the field of data governance and related fields who have taken the time to write articles and books so that I and many others could gain knowledge. The bibliography in this series shows the extent of my appreciation to those who have made that effort. Special thanks to Anne-Marie Smith, Boris Otto, Chisolm Malcolm, Carlo Batini, Dan Myers, Dan Power, Dannette McGilvray, David Loshin, David Plotkin, Doug Laney, Dylan Jones, Evan Levy, George Firican, Gwen Thomas, Hubert Österle, John Ladley, John Talburt, Jill Dyché, Kelle O'Neal, Larissa Moss, Larry P. English, Laura Sebastian-Coleman, Lowell Fryman, Majid Abai, Monica Scannapieco, Neera Bhansali, Nicola Askham, Peter Aiken, Philip Russom, Prashanth H. Southekal, R. D. McDowall, Robert F. Smallwood, Robert Seiner, Richard Wang, Sid Adelman, Steve Sarsfield, Sunil Soares, Thomas C. Redman, Todd Harbour, Tony Fisher, Wayne Eckerson, and Yinle Zhou.

I had quite a few questions about book publishing, and I am extremely grateful to Bill Hefley, Jill Dyché, Karl Wiegers, Laura Sebastian-Coleman, Nicole Radziwill, Sandeep Nagar, and Victor Squires for answering some of them. Many thanks to Satish Gawade for helping me with understanding the terms of the book contract.

Many thanks to Andres Perez, Christopher Butler, George Firican, Jill Dyché, John R. Talburt, John A. Zachman, Laura Sebastian-Coleman, Phil Watt, Shannon Fuller, Stan Rifkin, and Tony Epler for agreeing for an interview and sharing their unique perspectives.

I would like to thank the many clients and colleagues who have challenged and collaborated with me on so many initiatives over the years. I appreciate the opportunity to work with such high-quality people.

I am very grateful to Springer for giving me an opportunity to publish this book. I am particularly thankful to Anushangi Weerakoon for her continued cooperation and support for this project. She was extremely patient and flexible in accommodating my requests. Her unique perspective and feedback made this book so much better. I would like to thank Gowtham Chakravarthy and Saranya Kalidoss for making the production process smooth for me. Thanks to the Springer team for helping me make this book a reality. The Springer team made the process and the experience very easy and enjoyable. I am also thankful to the reviewers for their time and constructive feedback, that helped improve the quality of this book. Many thanks to Jeannine Siviy for her feedback and helpful suggestions that helped make this a better book.

I am grateful to my teachers at Sacred Heart Convent, DAV JVM, and Birla Institute of Technology, where I received the education that created opportunities that have led me to where I am today. Thanks to all my English teachers, and special thanks to Miss Amarjeet Singh, whose efforts sharpened my reading and writing skills. My years in Ph.D. research have played a key role in my career and personal development, and I owe a special thanks to my Ph.D. guides: Dr. Vandana Bhattacherjee, the late Dr. S. K. Mukherjee, and my teacher and mentor, Dr. P. K. Mahanti, who supported me during this period. Though miles away, Dr. Vandana Bhattacherjee and Dr. P. K. Mahanti still provide me with guidance and encouragement, and I will always be indebted to them. I am also thankful to my students whose questions have enabled me to think more and find a better solution.

Last but not least, many thanks to my parents for their unwavering support, encouragement, and optimism. They have been my rock throughout my life, even when they are not near me, and hence share credit for every goal I achieve. Writing this book took most of my time outside of work hours. I would not have been able to write the manuscript without their encouragement and support. They were my inspiration and fueled my determination to finish this book.

About This Book

Data is no longer a by-product of an organization's IT systems and applications, but is an organization's most valuable asset and resource, and has a real, measurable value. Companies that govern and manage their data effectively are able to achieve a competitive advantage in the marketplace. **Data Governance and Compliance: Evolving to our Current High Stakes Environment** is the first book in the mini-series, **Data Governance: The Way Forward** consisting of three books. This book sets the stage in terms of evolution of corporate governance, laws and regulations, other forms of governance, how data governance interacts with other corporate governance subdisciplines, and so on, and goes on to explain how data governance helps achieving compliance. This book shares the combined knowledge related to data and data governance that the author has gained over the years of working in different industrial and research programs and projects associated with data, processes, and technologies and unique perspectives of thought leaders and data experts through interviews conducted. This book will be highly beneficial for IT students, academicians, information management, and business professionals and researchers to enhance their knowledge and get guidance on their own specific data governance projects from a compliance perspective. This book is technology agnostic and contains a balance of concepts and examples and illustrations making it easy for the readers to understand and relate to their own specific data projects.

Contents

1 Introduction to Governance, Corporate Governance, and Compliance 1
1.1 Introduction 1
1.2 Governance Versus Management 1
1.3 Corporate Governance and Its Subdisciplines—in a Nutshell 3
1.4 Corporate Governance, the Digital Age, Data and Compliance 5
1.5 What's Ahead? 7
References 7

2 Governance Evolution and Basic Concepts 9
2.1 Introduction 9
2.2 Governance 10
2.3 Corporate Governance 10
2.4 Evolution of Corporate Governance 12
2.4.1 Seventeenth Century to Nineteenth Century 13
2.4.2 Early 20th Century to 1980 14
2.4.3 1980s Onwards 17
2.5 Concluding Thoughts 43
References 43

3 Corporate Governance Subdisciplines, Data, and Data Governance 51
3.1 Corporate Governance Subdisciplines—Introduction 51
3.2 Corporate Governance—The Approach 53
3.2.1 Operational Governance 54
3.2.2 Financial Governance 55
3.2.3 Human Resource Governance 56
3.2.4 Risk Governance 57
3.2.5 Security Governance 58
3.2.6 IT Governance 59
3.2.7 Data and Data Governance 63

3.3 Data Governance—History and Evolution 65
3.4 Data Governance Definition 67
3.5 Corporate Governance and Data Governance 76
3.6 IT Governance Versus Data Governance 77
3.7 Governance, Risk, and Compliance (GRC) 79
3.8 Data Governance and GRC 81
3.9 Tying Data Governance with the Different Disciplines of Governance 81
3.10 Concluding Thoughts 83
References 85

4 Regulations 89
4.1 Introduction 89
4.2 Health Insurance Portability and Accountability Act (HIPAA) 90
4.2.1 HIPAA Privacy Rule 91
4.2.2 HIPAA Security Rule 91
4.2.3 HIPAA Breach Notification Rule 91
4.3 General Data Protection Regulation (GDPR) 92
4.4 Know Your Customer (KYC), Anti-Money Laundering and Counter-Terrorism Financing Act 2006 (AML/CTF Act) 97
4.5 Foreign Account Tax Compliance Act (FATCA) 99
4.6 Basel Committee on Banking Supervision's Standard Number 239 (BCBS 239) 100
4.7 Common Reporting (COREP) and Financial Reporting (FINREP) 105
4.7.1 Common Reporting (COREP) 106
4.7.2 Financial Reporting (FINREP) 106
4.8 Payment Card Industry Data Security Standard (PCI DSS) 106
4.9 Concluding Thoughts 108
References 108

5 Data Governance and Compliance 109
5.1 Data Governance, Regulations, and Compliance 109
5.2 Cost of Non-compliance 113
5.3 Data and Regulations—The Challenges 118
5.4 Compliance, Data Characteristics, and Activities 122
5.4.1 Data Privacy 123
5.4.2 Data Security 123
5.4.3 Data Accessibility 124
5.4.4 Data Completeness 125
5.4.5 Data Accuracy 125
5.4.6 Data Consistency 125
5.4.7 Data Integrity 126
5.4.8 Data Currency and Timeliness 126
5.4.9 Traceability 126

5.4.10 Data Granularity 127
5.4.11 Data Usage 127
5.4.12 Data Retention 127
5.5 Data Characteristics and Compliance—Interrelationships 128
5.6 How Does Data Governance Help with Being Compliant? 129
5.6.1 Improved Data Discovery And Data Lineage 131
5.6.2 Enhanced Data Privacy and Security 131
5.6.3 Improved Data Quality 132
5.6.4 Effective Risk Management 132
5.6.5 Accountabilities and Responsibilities 132
5.6.6 Policies and Processes 133
5.6.7 Metrics 133
5.6.8 Education, Training, and Change Management 133
5.7 Regulatory Compliance and Data Management 134
5.7.1 Define and Discover Data 142
5.7.2 Data Profiling and Assessment 143
5.7.3 Compliance Data Store—Data Mapping and Transformation 145
5.7.4 Logging 147
5.7.5 Data Standardization 147
5.8 Concluding Thoughts 149
References 150

6 Data Governance and Compliance: Concluding Thoughts and the Way Ahead 155
6.1 Corporate Governance—Paving the Way 155
6.2 Data Governance Subdiscipline 155
6.3 Regulations, Compliance, and Data Governance 156
6.4 The Way Forward 157

7 Data Governance Interviews 159
7.1 Andres Perez, Information Management Consultant and President, IRM Consulting, Ltd. Co. 160
7.2 Christopher Butler, CDO HSBC UK, HSBC 161
7.3 George Firican, Director, Data Governance & Business Intelligence @UBC DAE | Founder of www.lightsondata.com 163
7.4 Jill Dyché, Principal, Jill Dyché, LLC. 164
7.5 Dr. John R. Talburt, Acxiom Chair of Information Quality at the University of Arkansas at Little Rock, and Lead Consultant for Data Governance and Data Integration with Noetic Partners Inc. 166
7.6 John A. Zachman, Author of "The Framework for Enterprise Architecture" (The "Zachman Framework"), Zachman International 171

7.7 Laura Sebastian-Coleman, Data Quality Lead, Finance Data Governance Center of Excellence, CVS Health 172
7.8 Phil Watt, Director, Elait Australia 174
7.9 Shannon Fuller, Director of Governance Advisory Services, Gray Matter Analytics 176
7.10 Dr. Stan Rifkin, Director of R&D, Master Systems Inc. 177
7.11 Tony K. Epler, Chief Data Strategist, PricewaterhouseCoopers 177

Appendix A: Bibliography 181

Appendix B: Glossary of Terms 183

Index 185

About the Author

Rupa Mahanti is a Business and Information Management consultant with extensive and diversified consulting experience in different solution environments, industry sectors, and geographies (United States, United Kingdom, India, and Australia).She has expertise in different information management disciplines, business process improvement, regulatory reporting, and more. Her research interests include quality management, information management, software engineering, empirical study, environmental management, simulation and modeling and more. With a work experience that spans industry, academics, and research, Rupa has guided a doctoral dissertation, published a large number of research articles, and is the author of the book—*Data Quality: Dimensions, Measurement, Strategy, Management and Governance (Quality Press).* She is an Associate Editor with the journal Software Quality Professional and a reviewer for several international journals.

Acronyms and Abbreviations

ACCA	Association of Chartered Certified Accountants
ACRA	Accounting and Corporate Regulatory Authority
AIGL	American Insurance Group
ALCOA	Attributable, legible, contemporaneously recorded, original or a true copy, and accurate
ALI	American Law Institute
AML	Anti Money Laundering
ASX	Australian Stock Exchange
ASXCG	ASX Corporate Governance Council
AUSTRAC	Australian Transaction Reports and Analysis Centre
BCCI	Bank of Credit and Commerce International
BI	Business Intelligence
BoG	Bank of Ghana
BSA	Bank Secrecy Act
CBN	Central Bank of Nigeria
CCAR	Comprehensive Capital Analysis and Review
CDS	Credit Default Swaps
CECL	Current Expected Credit Loss
CEO	Chief Executive Officer
CFO	Chief Financial Officer
cGMP	Current good manufacturing practices
CGOC	Compliance, Governance and Oversight Council
CII	Confederation of Indian Industry
CIO	Chief Information Officer
CMA	Capital Markets Authority
CobiT	Control Objective for Information and Related Technologies
COREP	Common Reporting Framework
COSO	Committee of Sponsoring Organizations
CPO	Chief Privacy Officer
CRD	Capital Requirements Directive
CRM	Customer Relationship Management
CRR	Capital Requirements Regulation

DG Data Governance
DGO Data Governance Office
DPA Data Processing Authority
DPO Data Protection Officer
DW Data Warehouse
EBA European Banking Authority
EC European Commission
ePHI Electronic Protected Health Information
ERP Enterprise Resource Planning
ETL Extract, Transform, Load
EU European Union
FAIR Factor Analysis of Information Risk
FASB Financial Accounting Standards Board
FATCA Foreign Account Tax Compliance Act
FCIC Financial Crisis Inquiry Commission
FCPA Foreign and Corrupt Practices Act
FFI Foreign Financial Institutions
FI Financial Institutions
FinCEN Financial Crimes Enforcement Network
FINRA Financial Industry Regulatory Authority
FINREP Financial Reporting Framework
FLSA Fair Labor Standards Act
FNMA Federal National Mortgage Association
FRC Financial Reporting Council
FSA Financial Services Authority
GDPR Global Data Protection Regulation
GFC Global Financial Crisis
GRC Governance, Risk, and Compliance
HIPAA Health Insurance Portability and Accountability Act
HITECH Health Information Technology for Economic and Clinical Health
HMO Health Maintenance Organization
HR Human Resources
IAIS International Association of Insurance Supervisors
IAPP International Association of Privacy Professionals
ICE Imperial Chemical Industry
ICO Information Commissioner's Office
ICPAK Institute of Certified Public Accountants
IEEPA International Emergency Economic Powers Act
IFRS International Financial Reporting Standards
IHIS Integrated Health Information Systems
IMDP Identification of Medicinal Products
IOG Institute on Governance
IOSCO International Organization of Securities Commissions
ISA Internal Security Assessor
ISO International Organization of Standards

IT	Information Technology
ITIL	IT Infrastructure Library
ITS	Implementing Technical Standards
ITSM	IT Service Management
JCC	Japanese Commercial Code
KYC	Know Your Customer
LEI	Legal Entity Identifiers
MDM	Master Data Management
MiFID	Markets in Financial Instruments Directive
MSA	Monetary Authority of Singapore
NAICOM	National Insurance Commission
NCCG	National Code of Corporate Governance
NED	Non-Executive Directors
NIST	National Institute of Standards and Technology
NSE	Nairobi Stock Exchange
OCR	Office for Civil Rights
OECD	Organization for Economic Co-operation and Development
OTC	Over the Counter
PAN	Primary Account Number
PCI DSS	Payment Card Industry Data Security Standard
PENCON	National Pension Commission
PHI	Protected Health Information
PII	Personally Identifiable Information
QSA	Qualified Security Assessor
RDA	Risk Data Aggregation
RDARR	Risk Data Aggregation and Risk Reporting
SAQ	Self-Assessment Questionnaire
SEC	Securities and Exchange Commission
SGX	Singapore Exchange
SIC	Standard Industry Classification
SID	Singapore Institute of Directors
SIMEX	Singapore Monetary Exchange
SME	Subject Matter Expert
SOE	State-Owned Enterprise
SoX	Sarbanes Oxley
SSN	Social Security Number
TWEA	Trading with the Enemy Act
ValIT	Value from IT investments
XBRL	eXtensible Business Reporting Language

List of Figures

Fig. 1.1 Subdisciplines of corporate governance 4
Fig. 1.2 Data, business processes, and business performance 6
Fig. 1.3 Cycle of compliance and performance 6
Fig. 3.1 Subset disciplines of corporate governance 52
Fig. 3.2 Structures, process and relational mechanisms for implementing IT governance (adapted from De Haes and Van Grembergen 2005) 61
Fig. 3.3 Data Classification by data sensitivity and protection 64
Fig. 3.4 Possible residence of Data Governance (DG) 68
Fig. 3.5 Six men and the blind elephant 73
Fig. 3.6 People aspect of data governance 75
Fig. 3.7 Good quality data, effective control, and compliance 76
Fig. 3.8 GRC overview (adapted from Salido et al. 2010) 80
Fig. 3.9 Tying the different disciplines of governance (adapted from Parkinson February 2016) 82
Fig. 3.10 Technology, widespread data access, and data governance 84
Fig. 4.1 GDPR principles 93
Fig. 4.2 Rights of individuals under GDPR 95
Fig. 4.3 BCBS 239 principles 100
Fig. 4.4 Overarching governance and infrastructure principles 101
Fig. 4.5 Risk data aggregation capabilities principles 102
Fig. 4.6 Risk reporting practices principles 103
Fig. 4.7 Supervisory review, tools and cooperation principles 104
Fig. 5.1 Compliance—setting the context 110
Fig. 5.2 Regulations and compliance 111
Fig. 5.3 Relationship between policies, behaviors, and compliance to rules, laws and regulations 112
Fig. 5.4 Data and regulations—challenges 119
Fig. 5.5 Compliance and data characteristics and activities 122
Fig. 5.6 Data governance and compliance—the five Ws and one H 129
Fig. 5.7 Role of data governance in compliance 130
Fig. 5.8 Regulatory compliance and data management 148

List of Tables

Table 3.1 Data governance definitions—literature review 70

Table 3.2 IT governance versus data governance 78

Table 4.1 Article 83 of GDPR—Two tiers of fines 92

Table 4.2 PCI DSS high level requirements (*Source* PCI SSC 2010) 107

Table 5.1 Some important regulations and the role of data governance ... 135

Table 6.1 Key milestones in the evolution of corporate governance, data, and data governance 156

Chapter 1
Introduction to Governance, Corporate Governance, and Compliance

> It is clear that good **corporate governance** makes good sense. The name of the game for a company in the 21st Century will be conform while it performs.
> —**Mervyn King** (Chairman: King Report).

Abstract This chapter introduces the readers to the concept of governance, corporate governance, the impact of the digital age and data on governance, compliance and performance, and discusses in detail the difference between management and governance.

1.1 Introduction

An entity is an organization, country, or business. At the core of any entity's success is good governance, which needs to be entrenched in its cultural fabric. Governance is the exercise of authority and control to ensure accountability and promote transparency. While good governance can lead to an entity's success, poor governance can result in its decline or failure.

Governance is often confused with management. The first section of this chapter discusses the differences between governance and management. The next sections introduce the readers to the concept of corporate governance, the impact of the digital age and data on governance, compliance, and performance.

1.2 Governance Versus Management

Governance and management are not the same. There is a thin line between governance and management, with both having leadership roles that need to be executed differently. Management focuses on people. In general, it concerns drive, commitment, allegiance, and politics. Governance focuses on power. It concerns policies, rules, regulations, the distribution of authority and responsibities, and the boundaries,

R. Mahanti, *Data Governance and Compliance*,
https://doi.org/10.1007/978-981-33-6877-4_1

exercise, and abuse of authority. Governance deals with "What," that is, determining the mission and vision, setting policies and strategic plans, selection of the chief executive, managing the governance process, providing insight, wisdom, and judgement, and verify, and audit performance.

Management focuses on operations and performance. In general, it concerns work done by people or machines and grouped into tasks, functions, or processes. Management deals with "How;" that is, implementing policies, strategic plans, and board decisions, supporting governance processes, establishing operating procedures, measuring performance, and delivering services (Community Southwark).

Bird (2001) states that while executives and managers administer, develop, implement and monitor business strategies on a day-to-day basis, boards and other governance structures deal with overall organization policy, culture, and direction. He goes on to state that, "Executives…manage organizations by the authority delegated to them by those who govern" (Bird 2001), thus making a strong distinction between the different elements of corporate governance and management (Webb et al. 2006). This distinction is based on the difference in the activities performed by the governance body, structures, and operational management.

However, the line between governance and management is not hard and fast and there are some overlap areas. The common overlap areas are ensuring that the strategic plan objectives are met, performance issues, and constant assessment of risk.

The distinction between governance and management differs from one organization to another organization and, from time to time, within the same organization, too (Mitchell 2010). The organizational maturity, size, and culture have an effect on the extent of separation between the governance and management structures.

Less mature organizations may take time to establish formal governance mechanisms. In the case of smaller organizations with limited employees or members, there might not be enough individuals to play distinct roles, and staff or members may perform multiple roles and may be entrusted with overlapping responsibilities of both governance and management. However, as an organization evolves and expands, the distinction becomes more and more important as delineating governance and management endorses accountability at all levels, and also provides a mechanism for good enterprise governance that emphasizes on stakeholder value by balancing performance and conformance (CommunityNet Aotearoa; Ball 2010; IEG World Bank 2007). Though a clean delineation is rarely possible, the 'right' balance between governance and management is necessary to avoid conflicts (Ball 2010).

The analogy of steering versus rowing is often used to describe the delineation of roles of governance versus management, and the delineation supports good decision making and outcomes driven activities. In the words of Robert Tricker, "If management is about running the business, governance is about seeing that it is run properly." While governance is about "doing the right thing," management is about "doing things right (Tricker 1998; IEG World Bank 2007)".

The governing body, such as a committee or board (comprised of directors), is responsible for governance and provides direction, leadership, control, and an enabling framework for collective decision making. At the same time, the management or executive team, led by the chief executive and his/her staff and volunteers, is responsible for the management and implementing decisions made by the governing body. The governing body's role is to oversee management and not to manage directly. It must be satisfied that the management team is doing its job under the stated policy and available resources (CommunityNet Aotearoa).

Every board is dissimilar in terms of how they carry out their governance role, and there should be clarity and consensus regarding the governance approach. According to Demb and Neubauer (1992), there are three main classic ways for boards to implement their governance role (Ball 2010);

- the watchdog,
- the trustee, and
- the pilot mode.

In a 'watchdog' role, theboard provides complete oversight and has no obvious involvement in the organization's activities. In the 'trustee' role, the board conducts itself like a custodian of assets and is accountable to shareholders and society for those assets. In a 'pilot' role, the board takes an active role in supervising the business of the organization (Demb and Neubauer 1992; Ball 2010).

As highlighted by John A. Zachman, Author of "The Framework for Enterprise Architecture" (The "Zachman Framework"), Zachman International, in his interview statement,—"the definition of "Management" historically included: 1. Plan 2. Organize 3. Staff 4. Direct, and 5. Control." Governance also includes almost all of these elements but with a different perspective. Governance systems include the management processes designed to deliver on performance objectives while considering stakeholder interests (ASQ).

1.3 Corporate Governance and Its Subdisciplines—in a Nutshell

Corporate governance is the system of policies, rules, mechanisms, practices, authorities, and processes used to direct, manage, and control a corporation. There are several subdisciplines under the corporate governance umbrella as shown in Fig. 1.1:

- Operational governance
- Financial governance
- HR governance
- Risk governance
- Security governance
- IT governance
- Data governance

IT governance and data governance are relatively new fields of governance as compared to the other corporate governance subdisciplines.

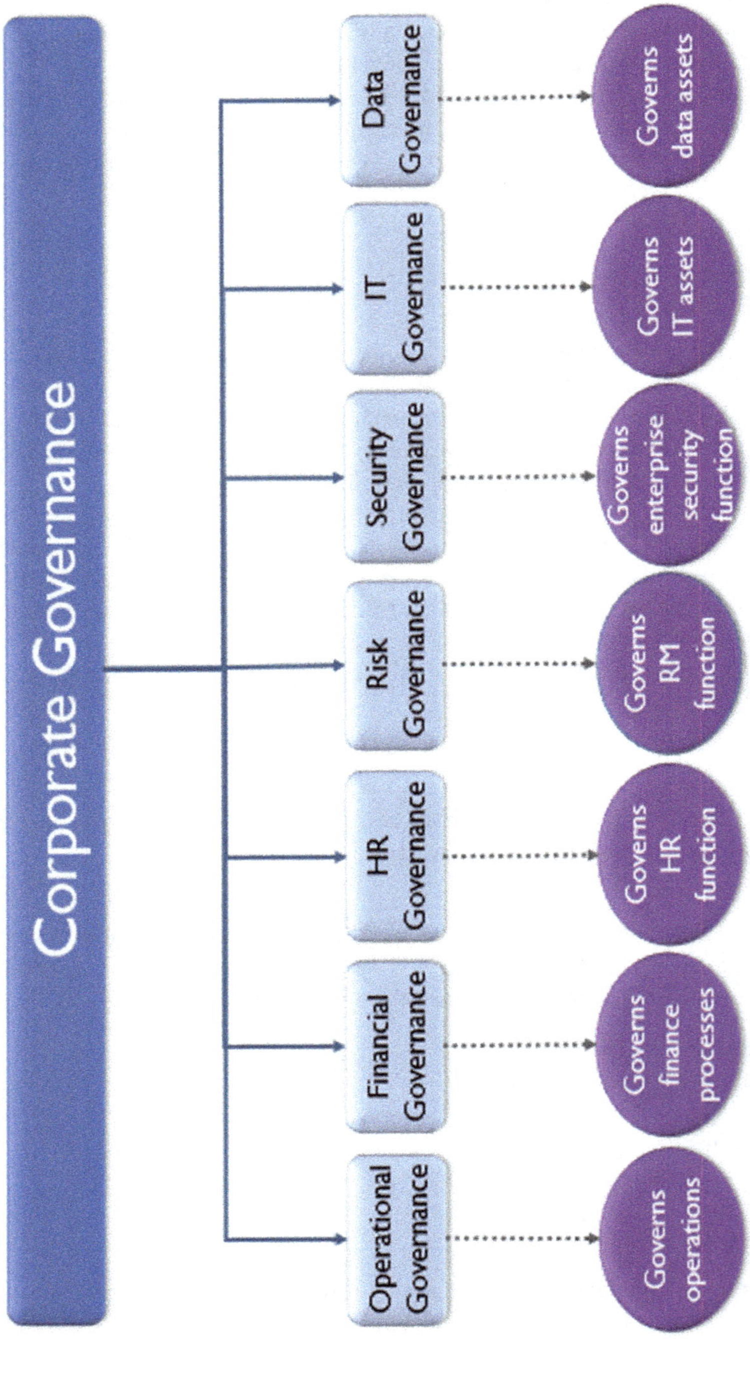

Fig. 1.1 Subdisciplines of corporate governance

All the subdisciplines of coporate governance have been discussed in detail in Chap. 3.

1.4 Corporate Governance, the Digital Age, Data and Compliance

> "We are governed by information and our belief in and acceptance of that information. Transparent, verified information and knowledge bridges are the most important tools available to us today." — Heather Marsh, The Creation of Me, Them and Us

Corporations have evolved over centuries, and so has the governance of corporations. Also, economic disasters, corporate scandals, and cases of fraud have weakened the world economy and influenced the evolution of governance. These have led to new laws and regulations to prevent recurrence of such incidents, events, or situations, which in turn have led to tighter controls, well-defined accountabilities, and enforcements to ensure compliance.

Effective governance needs facts, and good quality data represents facts. Before the advent of information technology, important facts were recorded on paper, kept under lock and key, and these served as inputs to the other disciplines of corporate governance.

The advancement of information technology, decreased cost of hardware and storage, and increased storage capacity has led us into the digital age. The digital age is characterized by organizations collecting large amounts of and different varieties of data electronically in shorter periods. In the digital age, data itself is an asset that needs to be effectively governed to ensure compliance and high-quality data. Compliance is one of the biggest drivers of data governance.

Data used in business processes has a direct or indirect impact on business performance. Faulty business processes often generate bad data, which adversely impacts business processes, which in turn adversely impacts business performance, as shown in Fig. 1.2.

Compliance is about following rules and regulations. These rules and regulations came into existence because of wrong actions, such as fraud and security breaches, as well as operational failures, corporate failures, and financial scandals that have been the result of poor performance or ethical issues. While fulfilling compliance obligations, organizations should look at improving business processes which will have a positive impact on performance, instead of only focusing on avoiding wrong actions. Where compliance hinders performance, the focus should be to achievecompliance by balancing performance.

Corporate failures, operational failures, security breaches, ethical issues and financial scandals beget rules and regulations which in turn beget governance, which begets improved performance and compliance, which in turn prevent corporate failures,

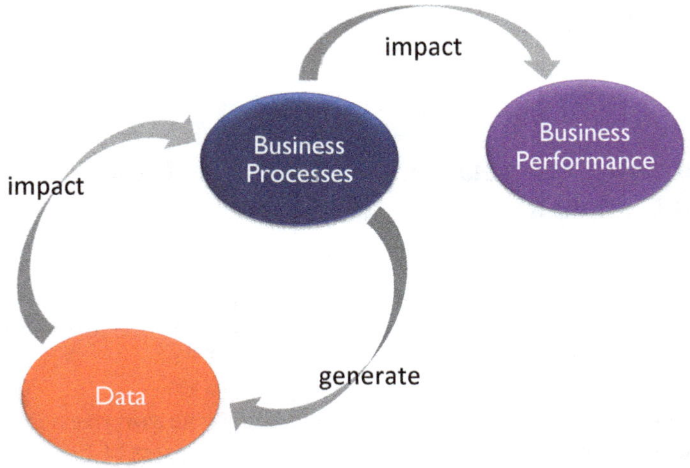

Fig. 1.2 Data, business processes, and business performance

operational failures, fraud, security breaches, ethical issues, and financial scandals as shown in Fig. 1.3.

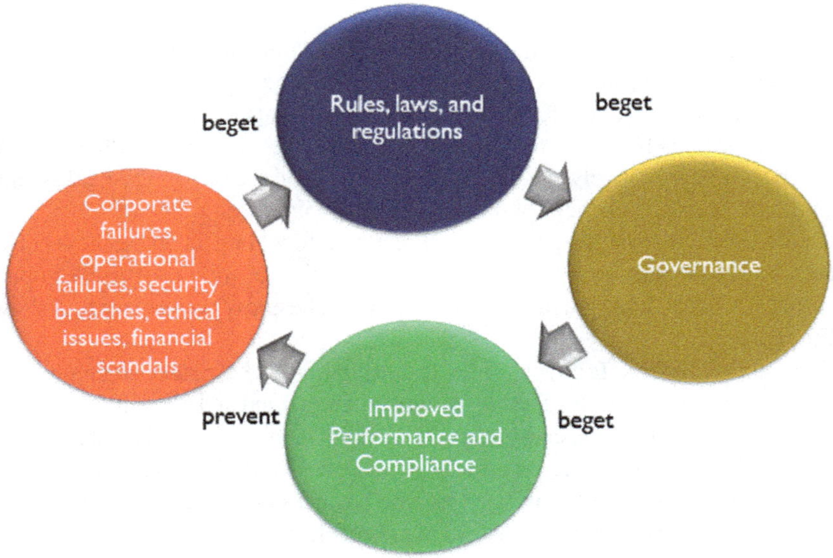

Fig. 1.3 Cycle of compliance and performance

1.5 What's Ahead?

In this book, we introduce the readers to the concept of governance, corporate governance, its history and evolution across different geographies in the face of scandals and financial disasters, which triggered the need for laws and regulations that organizations need to comply with, the different subdisciplines of corporate governance, and how the different subdisciplines of corporate governance tie together. We also will introduce the concept of data, data governance, the lack of standard definition of data governance, the distinction between IT governance and data governance, and finally, how data governance helps in attaining compliance.

References

ASQ. What is organizational or Corporate Governance, Quality Resources. https://asq.org/quality-resources/governance Accessed 20 Feb 2020

Ball D (2010) November 3, 2010, Management versus governance—it's not that easy, better boards. https://betterboards.net/governance/management-vs-governance/. Accessed 20 Feb 2020

Bird F (2001) Good governance: a philosophical discussion of the responsibilities and practices of organizational governors. Canadian J Admin Stud 298–312

CommunityNet Aotearoa. Governance and Management, NZ Navigator Trust. https://community.net.nz/resources/community-resource-kit/4-2-governance-governance-and-management/ Accessed 20 Feb 2020

Community Southwark (2018) Governance or Management: Knowing the difference. https://communitysouthwark.org/sites/default/files/images/Governance%20or%20Management.pdf. Accessed 20 Feb 2020

Demb A, Neubauer F (1992) The corporate board: confronting the paradoxes. Oxford University Press, New York

Independent Evaluation Group IEG World Bank (2007) Sourcebook for Evaluating Global and Regional Partnership Programs Indicative Principles and Standards. http://www.oecd.org/development/evaluation/dcdndep/37981082.pdf. Accessed 20 Feb 2020

Mitchell R (2010) The crucial difference between governance and management. AKT LTP. https://www.inphilanthropy.org/sites/default/files/resources/Crucial%20Difference%20Between%20Governance%20%26%20Management-AKT%20LLP-2011.pdf. Accessed 20 Feb 2020

Tricker R (1998) The role of management is to run the enterprise and that of the board is to see that it is being run well and in the right direction. In: Robert IT (ed) Pocket Director, p 8

Webb P, Pollard C, Ridley G (2006) Attempting to define IT governance: wisdom or folly? Proc Ann Hawaii Int Conf Syst Sci 8:194a–194a. https://doi.org/10.1109/HICSS.2006.68

Chapter 2
Governance Evolution and Basic Concepts

Necessity is the mother of invention.—Plato
The only constant in life is change.— Heraclitus of Ephesus

Abstract The concept of governance is the exercise of authority and control to ensure accountability and promote transparency. It has been around in some form since ancient times and has evolved with the requirements of the respective periods. This chapter explains governance and corporate governance followed by the evolution of corporate governance in the different geographies across the world.

2.1 Introduction

The concept of governance is the exercise of authority and control to ensure accountability and promote transparency. It has been around in some form since ancient times and has evolved with the requirements of the respective periods. In ancient times, governance centered around ruling tribes and kingdoms. Over time, corporations evolved, and a similar structure of governance was required. As corporations grew in size, and businesses, infrastructure, and technologies evolved, different subcategories of governance, such as operational governance, finance governance, risk governance, HR governance, security governance, IT governance, and data governance also evolved under the corporate governance umbrella to handle these changes and to ensure that organizations continued to run efficiently.

This chapter explains governance, corporate governance followed by the evolution of corporate governance in the different geographies across the world.

R. Mahanti, *Data Governance and Compliance*,
https://doi.org/10.1007/978-981-33-6877-4_2

2.2 Governance

Before we delve into the terminologies of corporate governance and data governance, let's start with the term "governance."

The concept of governance has been around for ages in the fields of politics and organizations, where it alludes to the task of running a government or an equivalent entity. The word "governance" can be traced back to the Latin verb "*gubernare*" and the Greek word "*kubernaein*," which means "to steer," and French words "*gouvernance*" and "*government*" whose meanings are very close, and allude to acts or manner of government.

The generic meaning of governance is the exercise of authority and control to ensure accountability and to enhance transparency or a method or system of government or management. Most definitions of governance rest on the following aspects:

- Authority
- Decision-making
- Responsibility and accountability

The Institute on Governance defines 'governance' as:

> The need for governance exists anytime a group of people come together to accomplish an end. Governance determines who has power, who makes decisions, how other players make their voice heard and how account is rendered. Governance is how society or groups within it, organize to make decisions. (IOG)

Business Dictionary defines 'governance' as:

> Establishment of policies, and continuous monitoring of their proper implementation, by the members of the governing body of an organization. It includes the mechanisms required to balance the powers of the members (with the associated accountability), and their primary duty of enhancing the prosperity and viability of the organization.

Both definitions of governance focus on authority, decision making, accountabilities, and responsibilities.

2.3 Corporate Governance

An organization consists of many stakeholders that include the Board, senior management, workforces, controlling authorities, investors and shareholders, clients, suppliers, government, and the society in which the organization operates. Effective collaboration with these stakeholders necessitates the organization to have sound corporate governance practices. These practices include the processes, policies, guidelines, principles, procedures, laws, rules, and regulations, which enable the stakeholders to interact in a transparent and workable fashion.

Corporate governance is the system of policies, rules, mechanisms, practices, authorities, and processes used to direct, manage, and control a corporation. The objective of corporate governance is to enable an active and practical management to deliver long term corporate and financial success.

Corporate governance is different from the day to day operational management of the company, which is carried out by full-time executives. It is a vital component in enhancing productivity, responsibility, and accountability, as well as in improving openness and transparency.

The Oxford University Press Business English Dictionary defines corporate governance as:

> the way in which directors and managers control a company and make decisions, especially decisions that have an important effect on the shareholders.

ICSA: The Governance Institute defines corporate governance as:

> the way in which companies are governed and to what purpose.

Generally, corporations have boards of directors who are responsible for the governance of their companies, and for improving transparency and accountability within the existing systems.

The Cadbury Committee in 1992 defines corporate governance as follows:

> Corporate governance is the system by which companies are directed and controlled.
>
> Boards of directors are responsible for the governance of their companies. The shareholders' role in governance is to appoint the directors and the auditors and to satisfy themselves that an appropriate governance structure is in place.
>
> The responsibilities of the board include setting the company's strategic aims, providing the leadership to put them into effect, supervising the management of the business and reporting to shareholders on their stewardship.
>
> The board's actions are subject to laws, regulations and the shareholders in general meeting.

The OECD Principles of Corporate Governance were developed in response to a request from an OECD Council Meeting at ministerial level in 1998 to produce a set of corporate governance standards and guidelines (Holt 2013). The OECD Principles of Corporate Governance (2004) provides one of the most used comprehensive definitions since it serves as a reference for all OECD member countries:

> corporate governance defines a set of relationships between a company's management, its board, its shareholders, and other stakeholders. Corporate governance also provides the structure through which the objectives of the company are set, and the means of attaining those objectives and monitoring performance are determined. Good corporate governance should provide proper incentives for the board and management to pursue objectives that are in the interests of the company and its shareholders and should facilitate effective monitoring. The presence of an effective corporate governance system, within an individual company and across an economy as a whole, helps to provide a degree of confidence that is necessary for the proper functioning of a market economy (Bank for International Settlements 2005).

The Australian Stock Exchange Corporate Governance Council defines corporate governance as (ASX 2003):

the system by which companies are directed and managed. It influences how the objectives of the company are set and achieved, how risk is monitored and assessed, and how performance is optimized.

Corporate governance became increasingly prevalent in the late twentieth century and in the early twenty-first century as a result of corporate scandals and company collapses due to operational failures, unethical practices, and exposures of accounting discrepancies and fraudulent practices.

2.4 Evolution of Corporate Governance

The roots of corporate governance lie in the term 'governance.' Governance is a very old concept, and can be traced back to ancient times. There has been some form of governance since mankind came into existence to sustain law and order in society. Tribal heads headed tribes, and kings and emperors ruled the world, and managed the administration of their state and subjects through effective governance.

For example, the Chinese Emperor Tang Taizong, a highly intelligent and ethical ruler, created a dynasty of prosperity and productivity that surpassed all other contemporary empires in culture, economy, agriculture, and transportation. Taizong ruled from 626 until 649, and his governance was deemed the Confucian ideal (Holt 2013).

Darius I of Persia (c.549–486/485 BC, Emperor of Persia 521–486/485 BC) was a proficient ruler who was known for his adept planning and organizational skills. He resolved outstanding wars, battles, and onslaughts, introduced a system of governance, started some large infrastructure projects, initiated and developed economic and trading alliances, and extended the empire overseas (Holt 2013).

In ancient India, the commencement of the Vedic age was characterized by tribes. The head of the tribe was the Raja or the King, though the concept of King had yet not developed. With the passage of time, large kingdoms started to grow, and by the sixth century BC, there were 16 Mahajanapadas (Kingdoms). The King was the absolute head of the legislative, executive, and judiciary branches. Several officials would assist the King in administering the state. The members of the council of minister would offer advice to the King, but final decisions were left to the King. The ministers and other officials were appointed by the King (Ancient Indian Government).

Chandragupta Maurya (reign: 321–298 BC), the founder of the Mauryan empire in ancient India, united and governed much of India under the direction of his teacher and chief minister, Chanakya. Chanakya was the author of the Arthashastra which is a text on economics, governance, and politics. He established a strong central administration from Pataliputra (now Patna in Bihar, India), in line with the Arthashastra (Chandragupta Maurya Wiki).

Ancient literature consisting of scriptures and religious texts allude to the existence of effective governance practiced during ancient times. For example, the Qur'an, which is the Holy Book of Islam, commands the faithful to follow the principles of shariah, which calls for ethical business behavior and sees money as a means for doing good. This is a guiding principle for the followers of Islam and can be found in current Islamic banking practices. There are also numerous passages in the Old Testament that caution against lending, corrupt, and unscrupulous behavior. China's Confucian philosophy calls on man to serve for the good of society and recognizes this as the highest calling (Tarantino 2008). Some Buddhist scriptures throw light on how the King used to manage his subjects.

In Indian scriptures and religious texts, good governance is often called Raj Dharma. That is righteous duty of the King (Kaushik). Ancient Indian scripturessuch as Shanti Parva-Anushasanparva of Mahabharat, Shukracharyas's Nitisar, Panini's Ashtadhyayi, Vedas, Aitreya Brahmana, Valmiki's Ramayana, and Kautilya's/Chanakya's Arthashastra also talk about good governance.

The Shanti Parva ("The Book of Peace"), which is the twelfth out of eighteen books of the Indian Epic Mahabharata, recites the duties of the ruler, dharma, and good governance counseled by the dying Bhishma and various sages (Mishra 2003). Bhishma, who had mastered the art of governance and had dedicated himself to the throne of his kingdom, Hastinapur, says: "the foundation for good governance is righteousness in public affairs. The King, his son—including relatives—his ministers, and the State employees who have taken the oath of their offices to uphold Dharma and to take care of the public needs, must not act unjustly or unethically because if they do so, they will not only destroy the moral basis of governance but will also turn the State into a hell" (Sharma 2003). Shanti Parva defines dharma not in terms of rituals or any religious precepts, but in terms of that which increases Satya (truth), Ahimsa (non-violence), Asteya (non-stealing of property created by another), Shoucham (purity), and Dama (restraint) (Suda 1970; Muniapan and Dass 2008). Chapter 109 of Shanti Parva asserts rulers have dharma (duty, responsibility) to help the upliftment of all living beings (Shanti Parva Wiki).

Chanakya's Arthashastra, written during the period of the Indian emperor, Chandragupta Maurya, in the fourth century BC, contains 15 chapters and 18 subchapters that include instructions about administration, management, law and justice, economy, and foreign policy. While highlighting the principle of good governance, it declares, "In the happiness of his people lies the King's happiness, in their welfare his welfare, whatever pleases himself he shall not consider as good, but whatever pleases his people he shall consider as good (Arthashastra; Kaushik)." This indicates, that the governance module proposed by Chanakya, is very similar to the corporate governance model practiced in today's world, wherein, we say that the directors should act in such a manner, that the stakeholders, investors, institutions, and others who work in the organization are benefited. According to Chanakya, good governance should avoid extreme decisions and extreme actions (Kaushik).

2.4.1 Seventeenth Century to Nineteenth Century

The contemporary practice of corporate governance has its origins in the 17th-century Dutch Republic (Frentrop 2003; De Jong et al. 2010; Gelderblom et al.

2010; Likomnik 2016). The first documented corporate governance dispute in history, (Mueller 2012) took place in 1609 between the shareholders/investors (most notably Isaac Le Maire) and the directors of the Dutch East India Company (VOC), which was the world's first formally listed public company (Funnell and Robertson 2013).

In 1794, Stewart Kyd defined a corporation as:

> A collection of many individuals united into one body, under a special denomination, having perpetual succession under an artificial form, and vested by the policy of the law with the capacity of acting in several respects as an individual.

In the nineteenth century, legislation around corporate governance and governance structures evolved in the form of the Chartered Companies Act 1837, Joint Stock Companies Act of 1844, Limited Liability Act of 1855, and the Companies Acts 1862–1893 in the United Kingdom. The origin of corporate governance can be tracked down to the formation of a registered company under the Joint Stock Companies Act of 1844 in the United Kingdom (Ruparelia and Njuguna 2016). This indicated the start of the modern corporation that divided control and ownership (Berle and Means 1967; Ruparelia and Njuguna 2016). These Acts together cover a variety of actions such as the governance of seals, stock, associations, as well as the registration and the formal closing of companies. The corporate governance frameworks continued to evolve to protect firms from the activities of professional managers with the passage of the Limited Liability Act of 1855 (UK) to protect shareholders from debt beyond their investment (Parker et al. 2002; Ruparelia and Njuguna 2016).

2.4.2 *Early 20th Century to 1980*

The concept of the modern corporation came into existence as a repercussion of the stock market crash of 1929 and the Great Depression of the 1930s that started in the United States but quickly extended to Europe and in due course to most parts of the globe (Tarantino 2008). The scars of the stock market crash of 1929 and the Great Depression of the 1930s left a mark ensuing generations and set the groundworks for government rules and corporate governance.

In 1932, Berle and Means, in their work, "The Modern Corporation and Private Property," emphasized the separation of ownership and control in large American corporations (Berle and Means 1932). In 1933, US President Franklin D. Roosevelt declared, "There must be a strict supervision of all banking and credits and investments; there must be an end to speculation with other people's money" (Rosenman 1938). The 1929 stock market crash resulted in the Congress passing two pieces of legislation:

Securities Act of 1933, and

Securities Exchange Act of 1934.

The above legislations serve as the foundation of U.S. securities laws. The Securities Act of 1933 was the first key part of federal legislation on the subject of the sale of securities. The Securities Exchange Act of 1934 focused on secondary markets, ensuring that exchanges, brokers, and dealers work in the top interests of investors. Provisions contained in the Act provided for the formation of the Securities Exchange Commission (SEC). The SEC was recognized as the primary regulator of US securities markets for regulating the markets and those who trade in those markets, a nonstop disclosure system for issuers and anti-fraud provisions (www.sia.com; Holton 2002).

The concept of "corporate government" is older than the concept of "corporate governance." While the concept of "corporate government" was already used in finance textbooks at the beginning of the twentieth century (Becht et al. 2004), the term "corporate governance" was first documented by Eells (1960, p. 108) to represent "the structure and functioning of the corporate polity." The idea of corporate governance advanced in 1962 to guarantee that investors get a reasonable return on their investment by providing suitable protection against misuse by management or inadequate utilization of their investment capital (Arsalidou and Wang 2005).

In the 1970s, the United States was the first country to use the term "corporate governance" consistently. This term became popular elsewhere a couple of decades later (Cheffins 2013). Hugo Young, a prominent U.K. political commentator, said of "governance" in a review of a 1976 book by the prominent Labour politician Harold Wilson, entitled "The Governance of Britain," that "it is a word devoid of modern meaning or use." Robert Tricker, in his 1984 book, *Corporate Governance: Practices, Procedures and Powers in British Companies,* observed, "Scant attention has been paid to governance in the British company" (Tricker 1984; Cheffins 2015). The fact that up to 1985 in UK, the term "corporate governance" had only ever been mentioned in the Times newspaper in a single 1978 article had propelled Tricker to make this observation (Cheffins 2015).

In the mid-1970s in the United States, corporate governance formed a part of the official reform agenda under the Securities and Exchange Commission (SEC) (Cheffins 2011). Penn Central, a railway that had expanded into pipelines, hotels, industrial parks, and commercial real estate went bankrupt in 1970 (Seligman 1982; Cheffins 2011). In 1974, proceedings were brought against three outside directors of Penn Central by the SEC, stating that they had changed the company's financial situation under federal securities law by failing to realize a extensive array of misconduct committed by the Penn Central executives (Schwartz 1976; Cheffins 2011).

By 1976, the year the term "corporate governance" first appeared in the Federal Register (Ocasio and Joseph 2005), the official journal of the federal government, the S.E.C., was beginning to treat managerial accountability issues as being part of its regulatory remit (Cheffins 2011).

Ralph Nader, Mark Green, and Joel Seligman, in their book, "*Taming the Giant Corporation*," published in 1976, supported a legal image of corporate governance with shareholders recommending directors who were authorized to manage the corporation. Directors, in turn, could delegate, as fitting, to executives serving as corporate officers (Nader et al. 1976; Cheffins 2011). Nader, Green, and Seligman deliberated that the legal model had little resemblance to the real-world truth of a dysfunctional "corporate autocracy," which was bent in favor of executives (Nader et al. 1976). They thought that corporate governance should be transformed through the endorsement of federal laws returning the board to its historical role as an internal auditor of the corporation liable for curbing the management from violations of law and breaches of trust (Nader et al. 1976; Cheffins 2011).

Their model for reformation had likeliness to the one presented by Melvin Eisenberg in his book on corporate law entitled "*The Structure of the Corporation: A Legal Analysis*," which was published in 1976. In his book, Melvin Eisenberg encouraged the replacement of the received legal model of the board with a "monitoring" model. In this model, a majority of directors would be unconnected to the management and the board's principal task would be to nurture managerial responsibility and liability (Wright et al. 2013; Eisenberg 1976; Cheffins 2011).

In the same year, an American Bar Association corporate law sub-committee, issued a Corporate Director's Guidebook that endorsed that significant outside director representation should be present on boards and that executive directors should be omitted from activities such as audit, compensation, and nominating committees boards (Subcommittee on the Functions and Responsibilities of Directors 1976; Small 2011; Cheffins 2011; Wright et al. 2013).

US Investigations into the Watergate scandal during the Nixon presidency revealed that the regulators and legislative bodies failed to control and stop several major corporations from making illegal political contributions and bribing government officials (Martin 2004). The SEC-sponsored voluntary disclosure program in the 1970s exposed that questionable or illegal payments above $300 million had been made to foreign government officials, politicians, and political parties by more than 450 U.S. companies (GAO Report 1981; Darrough 2004). Lockheed Martin Aircraft Corporation was found to have paid around $202 million in bribes in the U.S. as well as foreign countries including the Netherlands, Japan, Italy, Germany, Mexico, Spain, and Greece, in addition to contributing to the Nixon campaign (Darrough 2004).

The need to stop such corrupt propensities resulted in the legislation of the Foreign and Corrupt Practices Act (FCPA) of 1977 in America. FCPA initially barred U.S. corporations as well U.S. nationals from making inappropriate payments to foreign officers, parties or candidates, to help a company to acquire, retain or direct business to any person. The FCPA also imposed record-keeping and internal controls requirements on all companies subject to SEC dominion (Tarun 2006).

In 1977, the American Assembly issued a report which was considered by the New York Times as "the first draft of a new constitution for corporations" (New York Times 1977). This report provided the departure point for a 1978 symposium on "Corporate Governance in America" (Wright et al. 2013; Cheffins 2011).

In 1978, the Business Roundtable, a group which was established in 1974 to represent the views of 180 chief executive officers of major corporations, delivered a statement on "The Role and Composition of Directors of the Large Publicly Owned Corporation." This statement discussed the place of the board in "the corporate governance triad of shareowners/directors/operating management". It also recognized that the boards of public companies should characteristically have a composition with a majority of non-management directors and should establish audit, compensation and nomination committees dominated by outside directors (Wright et al. 2013; Business Roundtable 1978; Cheffins 2011).

The American Law Institute (ALI), a private organization which comprised of judges, practicing lawyers, and academics, dedicated itself in principle in 1978 to take on a project on corporate governance (Seligman 1987: 342–43; Small 2011: 135) and followed it up by establishing a conference in 1980. This conference was co-sponsored with the American Bar Association and the New York Stock Exchange and was attended by senior corporate executives, academics, lawyers and government officials. There was a general agreement in favor of the ALI's efforts at the conference, with the business community backing arising in part from a belief that ALI could provide, at least in comparison to legislative proposals Congress might come to consider—a controlled response to the events of the 1970s (Cheffins 2011).

However, there was a political shift, represented by Ronald Reagan's election to the presidency in 1980. This resulted in the end of the 1970s movement for corporate governance reform. The Protection of Shareholders' Rights Act of 1980 stalled in Congress. Likewise, the SEC seemed uninclined to pursue corporate governance reform with great vigor (Cheffins 2011).

2.4.3 1980s Onwards

In the 1980s, there were many corporate governance failures, which led to a number of nations establishing corporate governance codes in the early 1990s. Some of these were (Wiki1):

- Committee of Sponsoring Organizations of the Treadway Commission (USA),
- Cadbury Report (UK), and
- King Report (South Africa).

While corporate governance was in play before the 1980s, the eruption of corporate scandals in different parts of the world in the 1980s and the following decades such as the Enron scandal (USA), Tyco scandal (USA), Worldcom scandal (USA), Parmalat scandal (Italy), Polly Peck scandal (UK), Maxwell Communications Corporation scandal (UK), SwissAir scandal (Switzerland), Seimens scandal (Germany), UTI scandal (India), Satyam Scandal (India), Bannister scandal, (Dominic Republic), Olympus scandal (Japan), Sanlu Group scandal (China) and Urban Bank scandal (Philippines), and other factors like takeovers, mergers and acquisitions, stock market crashes, and globalization exposed the deficiencies in corporate governance and highlighted the need for increased regulations and reforms of corporate governance in organizations. All these factors have contributed to the constant evolution of corporate governance.

2.4.3.1 USA—1980s and 1990s

In 1985, a series of high profile business failures shook the United States. Over 1,000 savings and loan institutions were closed, holding over $500 billion in assets and representing about half of the total number of savings and loans. Deregulation, changing market conditions, volatile interest rates, tax changes, and reduced regulatory capital have all been cited as causes of the crisis. According to Timothy Curry and Lynn Shibut, losses totaled over $80 billion, with public sector/taxpayer costs of $75 billion and private sector costs of $7 billion (Curry and Shibut 1999; Tarantino 2008).

The US government appointed the Treadway Commission to determine the key reasons for misrepresentation in financial reports and to endorse ways of reducing such occurrences. The Treadway Commission published its report in 1987. The ***Treadway report*** highlighted the need for an appropriate control environment, independent audit committees, and an unprejudiced Internal Audit System. It stressed the need for published reports on the effectiveness of internal control. The Treadway Commission also advised the sponsoring organizations to develop an integrated set of internal control criteria to enable corporations to improve their control mechanisms. As a result of this recommendation, the Committee of Sponsoring Organizations (COSO) came into existence, which published a report in 1992. The COSO Report stipulated a control framework for the systematic functioning of corporations.

2.4.3.2 UK—1980s Onwards

In the late 1980s and early 1990s, a spate of corporate scams, scandals, and financial collapses surfaced in the UK due to poor management and lack of an effective control mechanism. It led to financial institutions and investors being concerned about their investments. It later prompted the UK government to recognize the ineffectiveness of the prevailing legislation and self-regulation as a measure of preventing scandals and financial collapses. For example, well-known corporations such as Polly Peck, Bank of Credit and Commerce International (BCCI), British & Commonwealth, and Robert Maxwell's Mirror Group International collapsed (Fernando 2011).

These collapses led to concerns around flawed financial reporting by companies. These failures resulted in the creation of several committees in the subsequent years to address corporate governance issues and prevent recurrence of such failures by improving corporate governance in the UK listed companies, with the first being the Cadbury Committee in May 1991.

The Financial Reporting Council, London Stock Exchange, and the professional accountancy institutions set up the ***Cadbury Committee*** in May 1991, to evaluate "The Financial Aspects of Corporate Governance" and to help raise the standards of British corporate governance and restore confidence in financial reporting and auditing. The Chairman of the committee was Sir Adrian Cadbury, chairman of the Imperial Chemical Industry (ICI), and the recommendations for good corporate governance were issued in the form of ***"The Cadbury Report"*** in 1992 which drew its name from Sir Cadbury (TCFACG-GEE 1992).

The Cadbury committee outlined the responsibilities of executive and independent directors, board audit committees and company accountants, and responsibilities of auditors to the shareholders and the board. The committee submitted its report and the "***Code of Best Practices***" in December 1992, to be practiced by listed companies. Many of the recommendations in the "***Code of Best Practices***" were already considered as best practice; however, they were not comprehensively applied. The Code endeavored to force companies, particularly public limited companies, to improve their conduct. One of the recommendations of the committee was that all the listed companies at the London Stock Exchange should embrace and conform to the best practices as outlined in the Code (TCFACG-GEE 1992).

The Code would be voluntary, but Companies were requested to release a statement of conformance/non-conformance with the Code of Best Practice in their annual financial statements including, the extent to which they had conformed to it and the reasons for any non-conformance. It was also advocated that the external auditors should be asked to assess the statement of conformance or non-conformance as part of their annual report to shareholders. The expectation was that the companies would follow guidelines that called for the institution of a sound governance structure, and a board of directors to monitor and govern the organization. One of the key recommendations in the report was the necessity to increase the number of independent non-executive directors, segregate the posts and responsibilities of the Chief Executive Officer of the company and the Chairman of the Board, and set up audit board committees with the capability to monitor and judge the actions undertaken by the company management (TCFACG-GEE 1992).

The Cadbury Report not only had a substantial domestic impact but also had a significant impact on an international scale as well. The Cadbury Code thus qualified as "the world leader" with respect to corporate governance issues in the 1990s (Vinten 1998). The reasons behind the widespread influence of the Cadbury report was the novelty and effectiveness of the Code of Best Practice, the "comply or explain" obligation which required the listed companies to adhere wholly to the provisions in the Code of Best Practice or disclose and explain if any non-compliance had a significant impact on the public company behaviors, and the wave of corporate scandals that erupted around the same time when the Cadbury Committee and its report came into existence (Ruparelia and Njuguna 2016).

In 1995, the ***Greenbury committee*** was set up under the chairmanship of Sir Richard Greenbury, Chief Executive Officer of Mark and Spencer, to look into the corporate governance issues related to directors' remuneration, in response to public complaints and concerns raised by shareholders regarding the same, and to ascertain good governance practices for determining the directors' remuneration. The committee published the ***Greenbury Report*** in 1995, which was named after the committee's chair. The report contained a "Code of Best Practice on directors' remuneration," which was to act as a guiding framework for all listed companies in the UK to make decisions on the director's remuneration and to comply with the same (Ruparelia and Njuguna 2016).

The London Stock Exchange embraced the recommendations of the ***Greenbury Report*** and integrated its recommendations into the UK Listing Rules. The ***Greenbury Report*** recommended the creation of Remuneration Committees in the boards consisting totally of non-executive directors (NEDs) to determine the remuneration of executive directors to avoid the potential of interest caused when the executive directors' pay was set by himself or other executive colleagues. Accountability, transparency, and linkage to performance formed the key principles in regards to directors' remuneration (Ruparelia and Njuguna 2016).

The ***Greenbury Report*** also included comprehensive recommendations related to disclosures about the remuneration of individual directors, the publication and disclosure of the remuneration policies guiding the remuneration of executive and non-executive directors, and the policies guiding how long a director could serve in a company. It was required that the annual report and accounts disclose for each named director, the different components of remuneration, such as salaries and fees, annual bonuses, deferred bonuses, and compensation for loss of office, and these detailed disclosure related to remuneration be checked by external auditors. The Greenbury recommendations did not need shareholder endorsement of the general remuneration policy for directors, or shareholder endorsement of short-term incentive schemes (Ruparelia and Njuguna 2016).

Both the Cadbury and the Greenbury Committees were set up as a reaction to the widespread opinion on the weaknesses in the corporate governance structure prevailing in the United Kingdom; they devised regulations for resolving the corporate governance problems of financial reporting and directors' remuneration respectively. These committees did not attempt a reform of corporate governance in its entirety (Dignam 1998a).

In 1998, a committee was set up under the chairmanship of Sir Ronald Hampel, chairman of ICI, to continue to review and revise corporate governance practices in the UK, following the recommendations of the Cadbury Report (1991) and the Greenbury Report (1995). The Hampel committee was the first committee to deal with the whole spectrum of corporate governance (Dignam 1998b). The Hampel Committee suggested that the recommendations of all three committees, namely the Cadbury Committee, Greenbury Committee, and the Hampel Committee, should be incorporated into a single code of corporate governance. The Hampel Committee sought to define broad principles over strict regulations, also known as the 'box-ticking' approach proposed by the Cadbury and Greenbury committees. The emphasis on broad principles rather than detailed rules and regulations was published as the ***Combined Code*** in 1998, which followed the publication of the ***Hampel Report*** (wiki-Hampel).

The Combined Code was adopted by the London Exchange and included in the UK Listing rules (Wiki -UK-CGC). The Combined Code was based on principles rather than rules, and companies could disregard aspects of the Combined Code. However, companies were required to disclose non-compliance as well as state the reason for non-compliance, in a similar fashion as that in the Cadbury Report. The findings of the Hampel committee encompassed an assessment of the roles of executive directors, non-executive directors, and auditing committees and their part in corporate governance (Ruparelia and Njuguna 2016).

As a result of the assessment, a list of approximately twenty corporate governance principles were sequestered and promoted as being necessary for good corporate governance. These principles encompassed the specific role of directors, recommendations relating to board remuneration, shareholders' rights and roles, and measures for ensuring accountability and transparency. It also recommended the institution of clear roles for executive and non-executive directors and reiterated the segregation of the Chief Executive Officer and the Board Chairman. Finally, the committee recommended that the processes related to the appointment of directors, remuneration, and auditing should be driven mainly by independent directors (Ruparelia and Njuguna 2016).

The *"Internal Control: Guidance for Directors of Listed companies Incorporated,"* referred to as the ***Turnbull Report***, was produced in 1999 by the Internal control Working Party of the Institute of Chartered Accountants in England and Wales. It was chaired by Nigel Turnbull, Executive Director of The Rank Group plc, and gave additional guidance to listed companies on how to implement the provisions of the Combined Code concerning internal control. The report informed directors of their commitments under the Combined Code to maintaining sound "internal controls" in their companies or running rigorous audits and checks to ensure accurate financial reporting and detect fraud before it became an issue (wiki3). Revised guidance was issued in 2005. The report was succeeded by a further FRC guidance issued in September 2014 (wiki3).

In 2003, Derek Higgs was appointed by the UK government to review the roles and effectiveness of non-executive directors as non-executive directors played a crucial role on the board. The detailed terms of reference included assessment of the population of non-executive directors, the process of recruiting and appointing them, their accountabilities, independence and effectiveness, their relationships (both actual and potential) with institutional investors, and remuneration of non-executive directors. The review also emphasized the link between independent directors and the financial performance of listed firms (Ruparelia and Njuguna 2016).

The committee produced a final report on the "Review of the role and effectiveness of non-executive directors," popularly called the ***Higgs Report***, published in 2003. The Higgs report contained recommendations relating to the structure of the board, the role and other commitments of the chair, the role of the non-executive director, the recruitment and selection procedures to the board, orientation and professional growth of directors, board tenure and time commitment, compensation, resignation procedures, audit and compensation committees, board liability, and relationships with shareholders (Ruparelia and Njuguna 2016).

One of the major recommendations of the review was to establish a clear definition of an independent non-executive director (Solomon and Solomon 2004). The definition of independence was to take into consideration the relationships that may have an impact on the director's impartiality. Another recommendation was that at least fifty percent of the board members must be independent directors, excluding the chairman and that the appointment of directors should consider the shareholders' concerns. Another recommendation was to keep the offices of the Chief Executive Officer (CEO) and the board chairman separate. The report recommended the establishment of a nomination committee by the board that would be responsible for evaluating individual directors, the board, and sub-committees annually. Finally, the report recommended that the level of the director's remuneration should be appropriate and reasonable to attract highly qualified individuals (Mallin 2006; Ruparelia and Njuguna 2016).

The Financial Reporting Council established a committee in 2002, headed by Sir Robert Smith, to look into the roles and responsibilities of audit committees, to develop the existing guidance for audit committees, and to address the concerns relating to the independence of auditors in the wake of the collapse of the auditing firm Arthur Andersen and the Enron scandal in the United States in 2002. The committee published a report in 2003 on "Guidance on Audit Committees," which was known as the ***Smith Report***, named after the committee chair. The Smith Report was published around the same time as the Higgs Report. This was followed by the ***Tyson Report*** on the recruitment and development of non-executive directors commissioned by the DTI.

A ***Combined Code of Corporate Governance*** was generated in 2003, which superseded the 1998 Combined Code. The Combined Code of Corporate Governance of 2003 was based on the recommendations of The Cadbury Report (1991), The Greenbury Report (1995), The Hampel Report (1998), The Higgs Report (2003), and The Smith Report (2003).

The Combined Code of Corporate Governance contained the main principles, the supporting principles and the provisions (practical requirements), and also covered the corporate principles of corporate governance; the roles and responsibilities of the board, the Chairman, the CEO, independent directors; director's remuneration; the formation and roles of the audit and remuneration committees; guidelines on the recruitment, appointment, and professional development of non-executive directors; relations with shareholders; provisions on the design of performance-related remuneration and a summary of corporate governance disclosure requirements (Ruparelia and Njuguna 2016).

The Combined Code of Corporate Governance is reviewed by FRC regularly to ensure that it remains "fit for purpose" and continues to promote improvement in the quality of governance. The Code has undergone several revisions since 2003, and the latest version of the Code was released in 2018 (Ruparelia and Njuguna 2016).

In October 2008, the UK Government commissioned Lord Turner to assess the causes of the global financial crisis. The ***Turner Review***, which was issued in March 2009, provides recommendations on the redesign of regulation and supervisory method required to form a more robust banking system for the future. The Review also emphasized the improvements in the efficiency of internal risk management and corporate governance (Financial Services Authority 2009; Kingston City Group).

In February 2009, Sir David Walker, ex-City regulator, had been asked by the Prime Minister to review corporate governance in the UK banking industry in the wake of critical loss and failure throughout the banking system and to make recommendations for corrective action. To prevent corporate governance failings in the financial industry sector, the ***Walker Review*** was issued in November 2009. It recommends more well-defined pay and bonus structures for all high earners (HM Treasury 2009, Kingston City Group).

2.4.3.3 Europe—1990s Onwards

The advent of corporate governance in Europe started with the Cadbury Report and the Code published by the UK in the 1990s, with Netherlands, France, and most other European countries following suit- all of which adopted one or more codes. Germany was one of the last countries to follow this trend and only adopted a corporate governance code in 2002 (Lannoo and Khachaturyan 2003).

In 1995, a study conducted by Ernst & Young for the Commission concluded that the EU should set framework principles for corporate governance of large companies in a directive, which could be transposed into the individual national legal systems either by law or within the framework of a national "best practice" code. Some harmonization of voting rights would also be desirable to increase the proportionality between the subscribed capital and the voting rights held (European Commission 1995). In February 1997, the Commission started a consultation on the simplification of company law. At a conference in December 1997, the Commission proposed creating a Company Law Forum, with an aim to better prepare Community initiatives (Lannoo and Khachaturyan 2004).

Following the US corporate scandals at the beginning of the twenty-first century, the European Union also experienced a few wide-ranging scandals, including the *Parmalat* scandal in Italy in 2002–2003, a $1.2 billion fraud in *Royal Ahold,* the Dutch Retailing giant, an uproar over executive compensation at *Skandia*, the Swedish insurance company, and accusations of fraud at *Vivendi International* (Landler 2003). As a result of these scandals, the European Commission (EC) added corporate governance issues to its agenda and presented an "Action Plan on Corporate Governance" in May 2003 (European Commission 2003).

2.4.3.4 USA—Post 2000s

Corporate governance and its effectiveness in the United States became all the more important, particularly in the aftermath of the high-profile company scandals and the collapse of Enron and MCI WorldCom and many other companies.

The Enron scandal, which came to light in October 2001, is the most significant corporate collapse in the United States since the failure of many savings and loan banks during the 1980s. Enron Corp, a Houston-based energy company, collapsed into a colossal bankruptcy due to a painstakingly-planned accounting fraud made by its accounting firm, Arthur Andersen. Enron shares dropped from $90 to $0.50, which was a huge financial disaster, where thousands of employees and investors lost their savings with the company as it filed for an earnings restatement in October 2001 (Jason 2018; Keudell 2012).

Arthur Andersen LLP was one of the ‘Big Five Audit Firms’ before it is collapse. The Enron and WorldCom accounting scandals, which exposed the firm’s failure to discharge its duties as auditors of the companies, led to its collapse, which resulted in the loss of 85,000 jobs and corporate re-branding (Jason 2018; Opinion Front).

WorldCom, the US telecommunications giant, filed for Chap. 11 in 2002. However, an internal audit report showed that the company was using fraudulent accounting methods to hide its declining financial condition. WorldCom inflated assets by as much as $11 billion resulting in 30,000 lost jobs and $180 billion in losses for investors. WorldCom’s internal auditing department uncovered $3.8 billion in fraud (Balkhi 2014).

The Enron scandal in the year of 2001, which ultimately led to the bankruptcy of the Enron Corporation, and the closure of Arthur Andersen, one of the five largest audit and accountancy firms in the world, was the largest bankruptcy reorganization in the history of USA at that time and unquestionably, also the largest audit failure.

As a result of the Enron and WorldCom scandals, new regulations and legislations were sanctioned to improve the reliability of financial reporting for public companies. The **Sarbanes-Oxley Act** was introduced in 2002, to avoid a repeat of such incidents by enhancing the accountability of auditing firms, to stay unbiased and independent of their clients.

Sarbanes–Oxley Act (SOX)–2002

In 2002 Congress passed the Sarbanes-Oxley Act (popularly known as “SOX”), which made wide-ranging improvements in several aspects of corporate governance. This act mandates the most far-reaching changes that Congress had imposed on the corporate arena since the FCPA of 1977 and the SEC Act of the 1930s with an intent to stop future scandals from happening and reassure the investors by the creation of a Public Company Accounting Oversight Board, revision of auditor independence rules, revision of corporate governance standards, and a significant increase in criminal penalties for securities laws violation (Gallegos 2003; Altintaş 2010).

Sarbanes Oxley aims to enhance corporate governance through measures that will strengthen internal checks and balances and, ultimately, strengthen corporate accountability. Some of these measures included the need for Chief Executive Officers and Chief Financial Officers to certify financial reports, Executive Officers and Directors being prohibited from taking personal loans, faster reporting of insiders trades, ban on insider trades during pension fund blackout periods, civil penalties added to disgorgement funds for the relief of victims- additional disclosure, auditor independence, including complete bans on certain types of work, and pre-certification by the company’s Audit Committee of all other non-audit work (O’Conor 2005; Altintaş 2010).

Sarbanes-Oxley significantly increased the penalties for willful violations of the Securities and Exchange Act of 1934, including the accounting provisions of the FCPA (penalties for violations of the anti-bribery provisions of the FCPA were not affected by Sarbanes–Oxley). The maximum penalty for violations by individuals was increased to a fine of not more than $5 million (from $1 million) and/or up to 20 years imprisonment (from 10 years). Maximum fines for violations by a corporation was increased from $2.5 million to $25 million (WileyRein and Fielding 2004; Altintaş 2010).

Sarbanes Oxley requires companies to perform a risk assessment of current information security policies to establish the extent to which such policies need updating to support the integrity of corporate financial information. The following two sections of SOX have a compliance impact on data and data governance:

- Sarbanes–Oxley, Section 302: Disclosure controls—requires that the company's principal officers (generally the Chief Executive Officer and Chief Financial Officer) certify and approve the integrity of their company financial reports quarterly (Kuschnik 2008) and annually. Internal access controls should be developed, implemented, and reviewed for deficiencies, and strengthened if needed at regular intervals to guarantee that the financial reports are of high quality and free from manipulation.
- Sarbanes–Oxley, Section 404: Assessment of internal controls—requires management and external auditors to account for and report on internal controls. Access controls should be maintained, reviewed for deficiencies, and strengthened if needed and reported at regular intervals so that their effectiveness is ensured, and the data is protected from unauthorized manipulation.

In addition to the Sarbanes–Oxley Act, many other new guidelines such as the U.S. Patriot Act (2001), the Basel II (2004), and the Basel III (2010) regulations were implemented post-2000. The Dodd-Frank Wall Street Reform and Consumer Protection Act (2010) evolved in the aftermath of the 2008 financial crisis. All these acts were put into practice to provide for improved financial systems for the financial services industry. Other industries have followed suit to improve corporate governance in their companies (Turner and Weickgenannt 2013).

The Dodd-Frank Act of 2010

The Dodd–Frank Wall Street Reform and Consumer Protection Act or the Dodd-Frank Act was signed into the federal law in 2010 (Library of Congress, Wiki2). Enforcement of this act has resulted in an improvement in financial stability and consumer protection (Bailey et al. 2017; Wiki2), although there has been a debate about its effects on the economy (Financial Times 2017; Marshall 2015; Wiki2). The Act established the Consumer Financial Protection Bureau (CFPB), which from its inception to April 2017 had "returned almost $12 billion to 29 million consumers and imposed about $600 million in civil penalties." (consumerfinance.gov; Wiki2).

On June 8, 2017, the Financial CHOICE Act was passed by the Republican-led House, which, if enacted, would roll back many of the provisions of Dodd-Frank. In June 2017, the Senate was creating its own reform bill (Borak 2017; Bryan 2017; Wiki2).

On March 14, 2018, the US Senate passed a bill, 67 to 31, easing financial regulations and reducing oversight for banks with assets below $250 billion (nbcnews.com; yahoo.com; Borak and Barrett; Wiki2). The purpose of the Dodd-Frank Act was to promote the financial stability of the United States by improving accountability and transparency in the financial system. Standards were implemented by the law to minimize risk-taking, increase data transparency and improve the efficiency with which data is aggregated and reported to regulators (Tarmy 2017), with the focus being on reference data or contractual information related to different financial instruments, entity reference data, which is used for counterparty risk assessment; pricing data, and data related to portfolio holdings (Brunelli).

2.4.3.5 Asia

In Asian countries, the attention to corporate governance has been intermittent. However, in the wake of the 1997–1998 financial crisis and stock market crashes, the interest in corporate governance spiked up in the late 1990s, and corporate governance practices strengthened. Following the occurrence of the Asian currency crisis in 1997, the flow of capital from foreign investors suddenly stopped, which led to an extreme liquidity problem in local capital markets and also harmed the economy as a result of inadequate capital as well as investor apprehensions. Several institutional and policy weaknesses came into light following the crisis (Yousuf and Islam 2015). The role of corporate governance in Asia has become more crucial, with the spate of scandals involving some of Asia's top companies—Satyam Computer Services (India), Sanlu Group (China), and Olympus (Japan), to name a few, which were tainted with the supposed lack of transparency or accountability to stakeholders of the companies.

In general, the Asian model of corporate governance is characterized by the concentration of family, families, or states' ownership of enterprises; the presence and enforcement of more internal corporate governance mechanisms by family members or government, with a primary focus on effective protection of the interests of majority owners (insider shareholders) and comparatively less protection for minority investors. This corporate governance model is known as the family-controlled or government-owned model (Rezaee 2011).

The following subsections discuss the evolution of corporate governance in various Asian countries, namely, China, Japan, Hong Kong, India, and Singapore.

China

In China, during the period 1950–1984, State-Owned Enterprises (SOEs) were in operation, and China's SOE governance can also be designated as a state-owned and managed model (Tarantino 2008). The state had ownership of all SOE properties. The state had managerial powers too, and all the resources, including labor force and financial resources, were planned for and assigned duties by the government. However, this centrally planned economy inhibited the development of the private sector. Furthermore, the SOEs were deprived of economic and legal independence. The SOE executives were required to achieve the production plans of the government rather than augment profits. Therefore, the governance structure of SOEs was a fundamental part of the general governmental framework. Executives of SOEs were appointed and dismissed by the government and the SOEs in this period served the government, rather than being independent economic entities (Chen 2015; Tarantino 2008).

From 1984 to 1993, the transitional model of SOE governance, also known as the state-creditor's rights model at the early stage of reform, came into place and was in operation until the Chinese Corporate Law of 1993 was enacted (Tarantino 2008). The reform aimed to confer legal status on SOEs, so that they would have the complete responsibility for their own profits and losses. The government retained the property ownership and segregated ownership rights from management rights, thinking that it would result in increased competitiveness amongst the SOEs.

During this period, due to the State-Owned Industrial Enterprises Law, commonly known as the SOEs Law issued in 1988, the corporate governance structure of SOEs went through several improvements. First, the governance mechanism treated the managers of SOEs as legal representatives. It awarded them with full responsibility for the operation of the enterprises, which required each SOE to establish a management consulting committee to assist the managers in making decisions on important issues. Second, each SOE established a local organization of the Chinese Communist Party to supervise the implementation of the government's policies. The government implemented the third SOE reform in 1993 to establish a modern corporate system in line with the market-oriented economy (Chen 2015; Tarantino 2008).

The policymakers began to investigate the modern corporate model within the Western world in a search for possible efficient solutions. The China Regulatory Commission released a Code of Corporate Governance in 2002 to promote effective and broad governance for listed companies in mainlandChina. The elementary principles of governance, details of shareholder rights protection strategies, description of the basic code of conduct and professional ethics for directors, administrators and senior management of registered companies were contained in this code (Rezaee 2018).

Approximately 49% of large- and medium-sized Chinese SOEs reported that they had suffered a loss by 1999 (China Confederation of Enterprises 1999). In 2005, China promoted privatization for the SOE development by introducing more sweeping corporate governance mechanisms from different capital markets, in which it succeeded to some extent (Chen 2015).

Hong Kong

As a former British colony, Hong Kong's regulatory system is built on the British common law; however, its business practices and corporate governance system have an Asian as well as British influence. The first Companies Ordinance was passed by the local legislature in 1865 when Hong Kong was a British Colony, and was based on the basis of the 1862 Act in the United Kingdom. Several amendments have been made to the Companies Ordinance since then. The three Companies Ordinances in 1865, 1911, and 1932 followed parallel company law reforms within the United Kingdom. In 1973, the Second Report of the then Company Law Revision Committee recommended major reforms based on the UK Companies Act, 1948, which were implemented in Hong Kong by the Companies (Amendment) Ordinance 1984 (Yu et al. 2014).

The first formal corporate governance initiative was launched in 1992 when the Stock Exchange of Hong Kong introduced the Corporate Governance Project to improve and promote a higher standard of corporate governance for Hong Kong's issuers. In 1993, the Stock Exchange of Hong Kong issued its own Code of Best Practice as an appendix to the Listing Rules that served as a non-mandatory guideline for issuers to devise their own codes of board practices (Hermes EOS 2017). While this was a step towards good corporate governance practice, corporate governance was not treated with earnestness by the majority of companies in Hong Kong, until the 1997 Asian financial crisis. In 1998, the establishment of an audit committee was introduced as a voluntary best practice (similar to the UK and Australian approach) for companies listed on the Main Board (Standard and Poor's 2002).

Hong Kong's corporate governance system has been seen to evolve in a progressive fashion. The Stock Exchange of Hong Kong issued the Code on Corporate Governance Practices, also known as the Old Code, which came into effect in January 2005 and was applicable to all companies listed on the Hong Kong Stock Exchange. This code's philosophy was in line with the then Combined Code in the UK, with listed companies required to comply with the 'code provisions' or were required to explain in case they deviated from compliance; the 'recommended best practices' were for guidance only.

The Code's organization and contents were also similar to the Combined Code with explicit chapters on directors, remuneration of directors and senior management, accountability and audit, delegation by the board, and communication with shareholders. Segregation of accountabilities of the management of the board and the day-to-day management of the company's business was also required by the Code. The Old Code was modified, following a consultation and review exercise in 2010. It was renamed as the Corporate Governance Code and became effective on April 1st, 2012. The Code sets out a number of principles, code provisions, and endorsed best practices. In addition to updating the Old Code, changes were made to the Listing Rules in the areas of board diversity, which came into effect from September 1st, 2013 (Cumming and Johan 2018).

In 2006, a comprehensive exercise was undertaken to amend the Companies Ordinance, to reform Hong Kong's company law, and further improve Hong Kong's position as a major international business and financial center. The Companies Ordinance addresses several facets of corporate governance such as the appointment of corporate directors, fiduciary duties of directors, in exercising the level of care, skills, and diligence, the preparation and disclosure of a comprehensive directors' report, and rules to address the possible conflicts of interest by directors (Rezaee 2018). A new Companies Ordinance came into effect in March 2014, after the enactment of subsidiary legislation, to enhance corporate governance. In particular, the new Companies Ordinance introduces a statutory provision that clarifies and codifies directors' duty of care, skill, and diligence (Yu et al. 2014).

India

In India, company legislation has been the main instrument to promote corporate governance for a significant length of time. The Companies Act in 1956 was a consolidating legislation that significantly transformed the corporate management structure in India. Followed by the inclusion of the Bhabha Committee recommendations, the Companies Act legislated, among other things, the elimination of the system of managing agencies. This institution had served the country justly and well during the initial period of corporatization and industrialization but had fallen from grace as a result of misuse and negligence in its application as time went by (Fernando 2011).

The need for corporate governance in India was highlighted in the wake of scams that had become an annual feature ever since the government liberalized the economy in 1991. The real history of corporate governance in India can be traced back to 1992 when efforts were made to put in place a system recommended by the Cadbury Committee. Since then, there had been the Harshad Mehta scam, Ketan Parikh scam, the UTI scam, Vanishing Companies scam, Bhansali scam, etc. (Fernando 2011).

The Confederation of Indian Industry (CII) outlined a voluntary code of corporate governance for listed companies in 1998. This was followed by the recommendations of the Kumar Mangalam Birla Committee set up in 1999 by SEBI to mandate international corporate governance standards for listed companies. The Kumar Mangalam Birla Committee separated its recommendations into mandatory and non-mandatory.

Mandatory recommendations encompassed such issues as the composition of the board, appointment and structure of audit committees, remuneration of directors, board procedures, and additional information regarding management, discussion, and analysis as a part of the annual report, a revelation of directors' interest, shareholders' rights and the compliance level of corporate governance in the annual report. Its non-mandatory recommendations comprised of issues concerning the chairman of the board, establishing a remuneration committee, half-yearly information to the shareholders, usage of postal ballots for certain key decisions, the appointment of nominee directors, and obligations of institutional shareholders (Fernando 2011).

The 2013 Companies Act in India makes wide-ranging improvements to practically all areas affecting corporate governance. The improvements include increased requirements for independent directors, added director obligations, the introduction of shareholder class action suits, and the proposal of the expanded use of e-governance for various processes in organizations (Yousuf and Islam 2015).

Singapore

Post its independence in 1965, **Singapore** pursued to establish a governance framework within which their businesses could operate. In 1967, the Companies Act, which is the foundation of companies' legislation in Singapore, came into force. The Monetary Authority of Singapore (MAS) was established in the form of a statutory board in the year 1970. In 1973, the Singapore Stock Exchange (SSE) came into being, when Malaysia opted out of the currency interchangeability agreement. The Securities Act was also enacted in 1973. The Pan El crisis of 1985 led to modifications in the Companies act in 1989, with audit committees and systems of internal accounting controls becoming compulsory for public companies and their subsidiaries. The Barring Bank collapse in 1995 resulted in extensive changes being made to the Future Trading Act (Koh and Yip 2015). In 1997, MAS carried out a major strategic assessment of the Singapore financial sector, as a part of its strategy to be a key financial center.

In the same year, the Investment Management Association of Singapore was established to commander the development and advancement of the investment and fund management industry in the country. As a result of the Asian Financial crisis in the late 1990s and following the MAS strategic assessment, several private sector-led committees (one of them being the Singapore Institute of Directors (SID)) were formed for evaluation of the characteristics of the prevailing systems. Their assessment and recommendations led to wide-ranging changes made to the Companies Act, and the institution of Singapore's first Code of Corporate Governance (Code) in 2001 (Koh and Yip 2015). The Singapore Code was based on the "comply or explain" principle first set out in the UK's Cadbury report (Koh and Yip 2015). The Singapore Institute of Directors has incorporated many initiatives to promote good corporate governance. It publishes "Statements of Good Practices" and has developed corporate governance guides for boards and major board committees.

In 1999, following the Exchanges Act, the Singapore Exchange (SGX) was formed from the merger of the SES, the Singapore Monetary Exchange (SIMEX) and the Securities Clearing and Computer Services Pvt. Ltd. The Securities and Futures Act was sanctioned in 2001, and SGX was given the twofold responsibility as a listed commercial exchange and a regulator/enforcer of listing rules (Koh and Yip). In 2004, the Accounting and Corporate Regulatory Authority (ACRA) was formed by merging the Registry of Companies and Business and Public Accountant's Board, to synergize the monitoring of corporate compliance and the regulation of public accountants performing a statutory audit.

Over the period of time, changes have been made to legislations and regulations in Singapore in order to respond to financial crises and to promote Singapore's vision of being a financial hub, and provide a socio-politically stable atmosphere, in which companies and financial institutions can function effectively. The Companies Act has undergone several iterations of substantial alterations in 2005, 2012, and 2014, some of which were influenced by the Sarbanes–Oxley Act and the Dodd Frank Act.

Japan

In Japan, the corporate governance system has developed over centuries, in the wake of changes, scandals, or crises. The traditional Japanese corporate governance model called "statutory auditor" (kansayaku) originated in the first Japanese Commercial Code (JCC) that was drafted in 1899 in general accordance with German corporate law (Ovsiannikov 2017). Kansayaku was initially a shareholder called to monitor the actions of the board of directors' and managers'. However, Kansayaku was not a part of an executive board (Benes 2013; Ovsiannikov 2017).

World War II marked a change in corporate governance. Preceding the militarization of Japan in the 1930s, there was a range of corporate governance patterns including joint-stock companies that were widely held, the state enterprise ownership and a limited partnership holding company structure, varied networks of publicly-quoted enterprises, that were controlled by family owners and operated by professional management known as *zaibasu* (Yao 2009; Jones and Zeitlin 2008; Morikawa 1992).

As Japan reconstructed its economy post-World War II, it established a distinctive corporate governance model, based on a combination of stable institutional ownership, main bank debt financing, professional and managerial control, and keiretsu transactions. The Japanese model of corporate governance was relatively different from the U.S. model of corporate governance. The main factors influencing the differences were the ownership structure, employment system, and relationship banking system.

The Japanese corporate governance model adequately sufficed the needs in the 1980s. The Japanese firms exhibited characteristics such as dispersed ownership, silent shareholders, and managerial control; and steered Japan in becoming the second-largest economy in the world by the 1980s. However, it struggled in the wake of economic crisis, scandals, and globalization in the 1990s. It was not able to service the needs in the 1990s. This led to subsequent reforms in corporate governance (Colpana et al. 2007).

As the financial performance of Japanese firms declined markedly from the early 1990s, the Japanese firms and government assessed and reformed their business and corporate governance practices, and began to implement some of the internal methods employed in US corporations, such as performance-based and incentive-ridden managerial compensation, as well as stock choices and more effective monitoring of board by shareholders. Also, their ownership structure changed to include more performance-oriented market investors (Colpana et al. 2007). There was a transition from a relationship-based, stakeholder-centered governance model towards a market-based and shareholder-oriented model to help Japan regain its global competitiveness. A series of modifications were made to the Commercial Code and other related laws throughout the 1990s and also into the twenty-first century.

The 2002 Commercial Code reform, which came into effect in April 2003, recommended large public companies "to allow outside directors to gain control of the board of directors through committees" (Itami 2005, p. 4) similar to that in the US, known as a three-committees' system. The 2002 Commercial Code contains institutional flexibility, which allowed choosing between statutory auditor and committee system (Ovsiannikov 2017).

Throughout the early 2000s, the government allowed the decision of whether to adopt the new scheme of corporate governance to be made by the listed companies. It was reflected in the implementation of the "comply or explain (the non-compliance)" principle by the largest stock exchange in Japan, Tokyo Stock Exchange (TSE) (2004, p. 23). The Tokyo Stock Exchange prepared the "Principles of Corporate Governance of Listed Companies," which was released in 2004, with an objective to provide the foundation for a common understanding of corporate governance by all the concerned parties, including listed companies and shareholders. The Companies Act of 2005 disconnected corporate law from the Commercial Code. The 2005 Act was followed by the securities law reform of 2006, which is known as the Japanese Sarbanes Oxley Act. In 2006, Nippon Keidanren (Japan Business Federation) prepared the guideline—"On the Desirable Corporate Governance System in Japan." (Nottage et al. 2008)

The guideline contemplates that the objective of firming corporate governance is to lay a groundwork that inhibits fraudulent company behavior while increasing profitability at the same time. The report emphasizes that enforcing a specific governance system is inappropriate and that it is desirable to consider the history and reality of Japan fully. Hence, it recommends that each company construct a useful framework that reflects its own vision. Massive reforms to Japanese corporate and commercial law, culminating in the consolidated Company Law (No. 86 of 2005: Takahashi and Shimizu 2005) and the 2006 amendments to securities regulations promised a more market-responsive and transparent model of corporate governance (Nottage et al. 2008).

Japan's Stewardship Code was introduced in 2014, which was followed by the introduction of the Corporate Governance Code in 2015. Both these codes influenced corporate behavior. They called for public companies and institutional investors to participate in productive discussions to improve shareholder value. The Stewardship Code contained principles modeled on the UK Stewardship Code and was based on Japan's current state of affairs (KPMG 2016). For the companies that opted to continue their listing at the Tokyo Stock Exchange, there have been no binding sanctions for the non-compliance with the 2002 Commercial Code requirements, mainly, the appointment of non-executives' majority to the board (Ovsiannikov 2017).

The requirements have been somehow severed recently, with the 2015 Japan's Corporate Governance Code (JCGC) (Ovsiannikov 2017) that proposes a new corporate governance structure for large public companies in Japan through an audit/supervisory committee, and new rules on outside directors. The new audit/supervisory committee governance structure was intended to facilitate appointments of outside directors and provide a corporate governance system that is more familiar to foreign investors. More than half of the statutory auditors must be from outside (Rezaee 2018). The amendments introduced in 2015 in Japan's Corporate Governance Code was intended to make Japanese corporate governance measures more consistent with those of the Western corporate governance (Rezaee 2018). The Stewardship Code underwent a revision in May 2017, which pushed for detailed proxy results disclosures (Nikko am).

2.4.3.6 Australia

In the first half of the twentieth century, Australian companies were best described as 'family capitalism' with important director level and managerial positions held by a close-knit business group (Ville and Merrett 2000; Fleming 2003). The nature of the separation of ownership from control changed in the second half of the twentieth century (Fleming 2003), with family capitalism being replaced by managerial capitalism, and institutional capitalism becoming more commonplace in the 1980s. In the 1960s, national corporations' legislation was encompassed in the Unified Companies Acts of the Australian states, policing of this regulation being a matter for companies' registrars as well as police fraud squads and the courts (Mees 2010).

The Australian regulatory system reforms in the 1970s and 1980s were not accomplished under the 'corporate governance' umbrella as the terminology was not used in Australia until the 1990s. The term 'corporate governance' was quite unknown in Australia in 1981 when the National Companies and Securities Commission (the predecessor to the Australian Securities and Investments Commission) first came into existence (Mees and Ramsay 2008). After the corporate scandals of the 1980s, the board rather than the company's managers became increasingly responsible for corporate governance matters, with Directors being held accountable for the success or failure of the company. The Directors were expected to monitor the management actions closely.

In Australia, the Bosch Report (Bosch 1991) had a major influence in the area of corporate governance reform (Kiel et al. 2012). The Bosch Report stressed accountability and transparency in board practices and on the role of the board in enhancing corporate performance. This report held the directors responsible for adequate management of the company to protect and improve shareholder value and to meet the company's commitments to shareholders, to the industry, and the law (ibid, Kiel et al. 2012). To attain these objectives, the report recommended the separation of the chair and CEO roles, board composition to have non-executive directors as a majority, and independence of at least one-third of the board. Also, it recommended that a statement of the company's main corporate governance practices be presented in each annual report (Kiel et al. 2012).

The Ramsay Report, published in 2001, on the independence of company auditors, discussed measures to enhance audit independence and recommended revisions to the Corporations Act as well as to the ASX Listing Rules. Australia experienced some high profile corporate failures, with HIH Insurance, One.Tel, and Ansett Airlines collapsing in 2001. These corporate failures lead to a call for more robust corporate governance practices. HIH Insurance, which had losses estimated between $3.6 billion and $5.3 billion, was the subject of a Royal Commission, which reported its findings in a three-volume report - *The Failure of HIH Insurance* in April 2003. The report of the HIH Royal Commission identified a number of possible breaches of the Corporations Act 2001 (Cth) (Corporations Act) and the Crimes Act 1914 (Cth) and made 61 policy recommendations on corporate governance, financial reporting, and assurance. These recommendations were considered in the finalization of the Australian Government's Corporate Law Economic Reform Program (CLERP 9) amendments to the Corporations Act, which dealt with continuous disclosure, including the introduction of personal liability for breach, auditor independence, accounting standards, expensing of options, compliance controls, and encouragement of greater shareholder participation at meetings (Kiel et al. 2012).

The Australian Stock Exchange (ASX) Corporate Governance Council came into existence in 2002 to devise and deliver an industry-wide corporate governance framework that could function as a practical guide for listed firms, their investors, the broader market, and the Australian community (ASX Corporate Governance Council 2003:Foreword). The Council comprised representatives from 21 stakeholder groups, including major investors, firms, and professions (see ASX Corporate Governance Council 2003:4), and delivered the first edition of the Australian Principles of Good Governance and Best Practices Recommendations in 2003 (ASX Corporate Governance Council 2003).

The first edition of the ASX Corporate Governance Council's (ASXCG) Principles of Good Corporate Governance and Best Practice Recommendations (ASX Principles) set out 10 "principles" of good governance (ASX Corporate Governance Council 2003). It was set up as a guideline. The ASXCG adopted the approach of recommending various corporate governance practices, requiring reporting on these practices as part of the corporate governance report under the ASX Listing Rules but at the same time giving companies the option to explain why they did not believe that a principle or recommendation is appropriate in their case.

The OECD revised its Principles of Corporate Governance Codes in 2004, and Australia also introduced revised codes in 2007. ASX Corporate Governance Council's (ASXCG) Principles of Good Corporate Governance and Best Practice Recommendations had undergone further revisions in 2010 and 2014.

In 2003, Standards Australia released a suite of standards, which includes AS 8000 Good Governance Principles, AS 8001 Fraud and Corruption Control, AS 8002 Organizational Codes of Conduct, AS 8003 Corporate Social Responsibility, and AS 8004 Whistleblower Protection Programs for Entities. Standards Australia's Good Governance Principles (AS 8000–2003) were developed to deliver a blueprint for the development and implementation of a generic governance system appropriate for different entities, including public and private companies, government entities, trustees, and not-for-profit organizations (Standards Australia 2003).

2.4.3.7 Africa

Until the 1980s, there were a high proportion of state-owned enterprises in all economic sectors in Africa. The issue of governance in Africa was first raised in 1988 in a World Bank report evaluating ten years of structural adjustment lending experience. The report identified serious institutional and managerial weaknesses in the public as well as the private sector (World Bank 1988; Kerandi 2008; Ruparelia and Njuguna 2016). The issue of "good governance" was further intensified by the 1989 World Bank report on Suburban South Africa (SSA) when the crisis in the region was labeled as a "crisis of governance" (World Bank 1989; Kerandi 2008; Ruparelia and Njuguna 2016). Economic globalization has been one of the major drivers for developing and adopting good corporate governance practices, which in turn, attract good investment. The following subsections discuss the evolution of corporate governance in various African countries, namely, South Africa, Nigeria, Ghana, and Kenya.

South Africa

Corporate governance in South Africa can be traced back to 1652. In 1652, Jan van Riebeeck, on behalf of the Dutch East India Company (DEIC), established the Cape of Good Hope. The laws governing companies during that time period was known as the Roman-Dutch law (Thompson 2001). In 1795, Great Britain took control of the Cape of Good Hope. After a few years, Great Britain returned the Cape of Good Hope to the Dutch. However, Great Britain again seized the Cape of Good Hope in 1876, shortly after the bankruptcy of the DEIC (Thompson 2001).

The British Government introduced English law, also known as the common and statutory law, in the nineteenth century. Company law has existed in South Africa since 1861, beginning with the Joint Stock Companies Limited Liabilities Act No 23 of 1861 of the Cape Colony. This Act, along with other provincial company legislation, was a carbon copy of the equivalent English legislation. The first national company law was introduced in 1926 with the Union Companies Act, which was amended from time to time along the lines of the latest English legislation. The 1926 Act was replaced in 1973 by the Companies Act No 61 of 1973, which, despite efforts to innovate and develop a direction more appropriate for South Africa, remains much in the mold of English law (Republic of South Africa 2004). Before corporate governance was institutionalized in South Africa, its requirements were based on the Companies Act, 61 of 1973, as well as the common law.

Following South Africa's re-admission to the global economy after its independence in the early 1990s, good corporate governance practices were recognized as a necessary element to operate with credibility in the international markets. In general, South Africa's corporate governance framework has developed largely in line with the Anglo-American style of corporate governance. In 1992, the King Committee was established under the patronage of the Institute of the Directors of Southern Africa, with the explicit objective of researching and making recommendations into corporate governance in South Africa.

The King Committee was chaired by retired Supreme Court of South Africa judge Mervyn E. King. In 1994, the committee issued the first King Report (named after its chair)—King I, and the first corporate governance code for South Africa. The King report was modeled on the Cadbury Report and is based on principles and practices. It also advocates an "apply or explain" approach. It was applicable to the following entities (Wiki-King Report):

- all companies listed on the main board of the Johannesburg Stock Exchange,
- large public entities as defined by the Public Entities Act of South Africa,
- banks, financial and insurance companies as defined by the Financial Services Acts of South Africa, and
- large unlisted companies.

Thekey principles covered by the first King report included the board of directors makeup and mandate, as well as the role of non-executive directors and guidance on the groups of individuals appropriate for non-executive directors, board appointments as well as guidance on executive directors' maximum term, determination, and disclosure of executive and non-executive director's remuneration, frequency of Board meetings, balanced annual reporting, the requirement for effective auditing, affirmative action programs, and the company's code of ethic (The Institute of Directors of Southern Africa 1994). Corporate governance in South Africa gained momentum after the 1994 King I report.

The changing global economic environment and legislative developments made it necessary to revise the 1994 King I report on corporate governance. The King Committee developed the King Report on Corporate Governance for South Africa, 2002 (***King II***), and the revised code of governance was applicable from March 2002. In addition to the types of organizations listed in King I, King II was applicable to departments of state, provincial or local government administration falling under the Local Government: Municipal Finance Management Act, and public institution or official exercising authority or performing a function in terms of the constitution, or exercising a public power or performing a public function in terms of any legislation, not including the law court or judicial officers. The key principles from the second King report were related to directors and their responsibility, risk management, internal audit, integrated sustainability reporting, accounting, and auditing (The Institute of Directors of Southern Africa 2002; Wiki-King Report).

The South African government passed the Corporate Laws Amendment Act in 2006 (Republic of South Africa 2006), which captures the most important changes to the Companies Amendment Act from 2004 (Republic of South Africa 2004), which consist of amendments of the Company's Act from 1973 (Republic of South Africa 1973). The Corporate Laws Amendment Act, 2006 (Republic of South Africa 2006) incorporates the recommendations made by the King II report (Institute of Directors Southern Africa 2002), thereby legalizing corporate governance in South Africa.

The new Companies Act no. 71 of 2008 ('the Act') and changes in international governance trends, necessitated the third report on corporate governance in South Africa, known as King III. The ***King III*** report was compiled by the King Committee with the help of the King subcommittees and was applicable from March 2010.

The report recommended that organizations generate an integrated report in place of an annual financial report and a separate sustainability report in accordance with the Global Reporting Initiative's Sustainability Reporting Guidelines (South African Institute of Chartered Accountants 2010; Wiki-King Report). The report incorporated a number of global, emerging governance trends, including alternative dispute resolution, risk-based internal audit, shareholder approval of non-executive directors' remuneration, evaluation of board and directors' performance (The Institute of Directors of Southern Africa 2009). It also incorporated several new principles to address elements that were not included in the previous King reports such as IT governance, business rescue, fundamental and affected transactions in terms of director's responsibilities during mergers, acquisitions, and amalgamations.

The corporate governance code is not imposed through legislation. However, owing to the progress in South African law, many of the principles put forward in King II are now personified as law in the Companies Act of South Africa of 2008. In line with the earlier versions, King III advocates an "apply or explain" approach. In contrast to the King I and King II, King III applies to all entities: public, private and non-profit (Wiki-King Report; Naspers).

The fourth revision of the Kings report, ***King IV***, was published by the King committee in 2016 and was effective from April 2017. It was necessitated by significant corporate governance and regulatory developments, which had taken place both locally and internationally since King III was issued in 2009. The other factor necessitating the revision was that while the listed companies were generally applying King III, non-profit organizations, private companies, and entities in the public sector were experiencing challenges in interpreting and adapting King III to their specific circumstances (Wiki-King Report; Institute of Directors in Southern Africa; Institute of Directors of Southern Africa 2016).

King IV was built on King III, and the enhancement was aimed at making King IV more accessible to all types of entities across sectors. In contrast to the earlier versions, which advocate an "Apply or Explain" approach, King IV requires an "Apply and Explain" approach to disclosure. King IV contains 16 principles with 208 recommended practices that apply to all organizations and a 17th principle under which are listed six recommended practices that apply to institutional investors. The key principles from the fourth King report were related leadership, organizational ethics, responsible corporate citizenship, strategy and performance, reporting, primary roles and responsibilities of the governing body, composition of the governing body, committees of the governing body, evaluation of the performance of the governing body, appointment and delegation to management, risk governance, technology and information governance, compliance governance, remuneration governance, assurance, stakeholders, and responsibilities of institutional investors (Institute of Directors of Southern Africa 2016).

Another reform in South Africa's corporate governance was the revision of the listings requirements by the JSE, which required companies to disclose minimum corporate governance statements in their annual reports. Other legislative developments since the publication of the King I report (The Institute of Directors of Southern Africa 1994) included the introduction of the Insider Trading Act, 1998 (Republic of South Africa 1998), which provides for more rigorous supervision and monitoring of insider trading, the Public Finance Management Act, 1999 (Republic of South Africa) which brings into force more stringent provisions for reporting and accountability by adopting an approach to financial management in government that focuses on outputs and responsibilities rather than the rule-driven approach under previous legislation. Finally, there has been a comprehensive update of the provisions and regulations governing the Banks Act, 1990 (Republic of South Africa), which substantially enforces higher levels of corporate governance compliance and risk reporting in banking institutions (Naidoo 2002; The Institute of Directors of Southern Africa 2002).

Kenya

Before liberalization of Kenya's economy in the 1990s, which institutionalized privatization of government corporations, there was no corporate governance framework. The establishment and subsequent inauguration of the Capital Markets Authority (CMA) in 1990 did not fundamentally alter the corporate governance landscape in Kenya (Ruparelia and Njuguna 2016).

Corporate governance in Kenya started gaining momentum in 1998, when a workshop on the role of non-executive directors was organized in Nairobi by M/s Dominion Consultants Limited, with a large number of leading corporate organizations participating in the workshop. Some of the factors driving the need for good corporate governance were privatization and corporatization of the economies on a global scale, corporate scandals and collapse of big corporations, and demand for better transparency and disclosure of information by shareholders. The workshop was sponsored and supported by major institutions which had a profound interest in corporate governance such as the Nairobi Stock Exchange (NSE), the Capital Markets Authority (CMA), the Institute of Certified Public Accountants (ICPAK), and the Kenya Chapter of the Association of Chartered Certified Accountants (ACCA) (PSICG).

While corporate governance and related issues were discussed to some extent, the workshop participants felt the need to convene again to discuss corporate governance in greater detail. To this end, a second seminar was held in March 1999 in Mombasa, in which decisions were made for an interim committee to frame a Code of Best Practice for Corporate Governance in Kenya, and to co-ordinate, where applicable, with other efforts to improve Corporate Governance and a Private Sector Initiative for Corporate Governance to be established to create a code of best practice for corporate governance in Kenya (PSICG).

A first draft code of best practice was created by the interim committee, which was circulated to over four hundred corporate organizations, development agencies, embassies, and government divisions with an appeal to review and provide feedback regarding the draft and the way forward. With the support of three development agencies, that is, the Ford Foundation, the British Department for International Development, and the Friedrich Ebert Foundation, a two-day workshop of specialists was held from 6–7th October 1999 to conduct further discussions and to arrive at a broader agreement. This was followed by a seminar organized by the Private Sector Initiative and attended by representatives from over seventy corporate and other organizations on 8th October 1999 (PSICG).

The Code of Best Practice for Corporate Governance, which was previously circulated was subsequently refined through expert input and comments from corporate respondents. Participants at the workshop and seminar concluded that the Code of Best Practice for Corporate Governance be adopted, printed, and distributed as a guide for Corporate Governance in Kenya. On October 8, 1999, the Corporate Sector at the seminar also mandated the Private Sector Initiative to establish the Corporate Sector Foundation, and work together with the Global Corporate Governance Forum, the Commonwealth Association for Corporate Governance, the African Capital Markets Forum, Uganda, and Tanzania in promoting good corporate governance (PSICG). The mission to institutionalize principles of corporate governance in Kenya concluded in the promulgation of the Guidelines on Principles of Corporate Governance for Public Listed Companies in 2002 (Du Plessis et al. 2005; Gakeri 2013).

Similar to the UK, Kenya also adopted non-statutory Guidelines and implemented the "Comply or Explain" approach. However, the "Comply or Explain" approach did not work as most companies never really complied with the 2002 Guidelines. Compliance was only a matter of "ticking off a box," and as a result, between 2002 and 2015, many Kenyan companies collapsed due to poor corporate governance (Mukabwa 2016).

With the aim of addressing the incompliance problem, CMA published the Code of Corporate Governance Practices for Issuers of Securities to the Public in 2015 (the 2015 Code), which succeeded the 2002 Guidelines. Through the 2015 Code, the Authority advocates the adoption of standards of governance that go beyond the minimum standards set in legislation, including in the new Companies Act, 2015. One of the major modifications in the 2015 Code is that it marked a shift from the "Comply or Explain" approach to the "Apply or Explain" approach, which required Boards to comply with the 2015 Code fully, and in case of failure to comply, the non-compliant companies must disclose to the CMA, the reasons for non-application and clearly indicate the time frame required and the strategies to be put in place towards full compliance (KCGPC 2015).

The "Apply or Explain" approach means that the Board could jointly conclude that implementing a recommendation would not, in specific situations, be suitable for the company. The recommendation could be implemented differently, or another practice might be applied while still realizing the objective of overarching corporate governance principles of fairness, accountability, responsibility, and transparency

(BG Attorneys). It must be noted, though, that there are some compulsory requirements that listed companies have no alternative but to implement these recommendations. These requirements were incorporated in the Capital Markets (Securities) (Public Offers, Listing, and Disclosures) (Amendment) Regulations, 2016 (Mukabwa 2016).

The Code of Governance for State Corporations (The Mwongozo) was developed by the Presidential Task Force and to incorporate principles of good corporate governance in the management of the State Corporations. Mwongozo adopted the "Comply or Explain" approach. Article 10 of the Constitution of Kenya anchors the Mwongozo, which outlines the national values and governance principles; Article 73 which requires that State Officers uphold public trust, honor, respect, and confidence; and Article 232 which lays down the values and principles of public service (Mukabwa 2016).

In July 2017, the Companies (General) (Amendment) Regulations, 2017, were published, which called for increased transparency concerning remuneration of directors.

Nigeria

Before 1960, Nigeria was a British colony. Post-independence, the UK legislation was reviewed, and the Companies Ordinances of 1922 was revoked and replaced by the Companies Act 1968, which was comprehensive legislation modeled after the 1948 Companies Act from the United Kingdom. Before 1990, the 1968 Companies Act was the principal Company Law statute in Nigeria and contained detailed provisions in relation to the running of companies in relation to the roles of the board of directors and the members in general meeting. The Companies Act, 1968, was revoked because of its limitations and criticism from stakeholders and was replaced in 1990 by the then Companies and Allied Matters Decree No, 1 of 1990. With some inconsequential amendments over the years, this statute which is now known as the Companies and Allied Matters Act, Cap. C20, Laws of the Federation of Nigeria 2004, is the principal statute regulating companies in Nigeria (Marshall 2015; Nat 2013).

The financial crisis in Nigeria prior to the 2000s, and the recognition that the basic company law did not address the corporate challenges at the time, highlighted the need for a formal governance code. Also, poor corporate governance at the time was recognized as one of the major factors contributing to the financial sector crisis. A formal corporate governance code, the Code of Corporate Governance for Banks and Other Financial Institutions in Nigeria, was issued by the Bankers' Committee in August 2003. The code was compiled by the Bankers' Committee's Sub-Committee on corporate governance. The 11 principles of the code focus on responsibilities of Board of Directors, the structure of Board of Directors, the Chairman and the Chief Executive Officer, appointments to the Board, Directors remuneration, Board performance and assessment, risk management, financial disclosure, relations to shareholders, and audit committee. The code applied to all banks and other financial institutions operating in Nigeria at the time of its issuance (Marshall 2015; Nat 2013).

However, the code was not popular, as it was issued by a voluntary organization and not a regulator, and failed to make an impact. Another reason contributing to the lack of impact was that the Securities and Exchange Commission (SEC) issued another legislation; just two months after the Bankers Committee had issued its code. In October 2003, SEC's 17-member committee, headed by Atedo Peterside, delivered the Code of Best Practices on Corporate Governance in Nigeria. The SEC 2003 code emphasized the role of the board of directors and management, shareholder rights and privileges, and the audit committee (Marshall 2015; ibid, Olutuyi 2017; Nat 2013).

While the SEC 2003 Code made major reforms, it still had a few shortcomings. In order to address these shortcomings, the following three industry-specific codes were issued (Marshall 2015):

- Code of Corporate Governance for Banks in Nigeria Post Consolidation issued by the Central Bank of Nigeria (CBN) in 2006, to ensure accountability on the part of bank CEOs.
- The 2008 PENCOM Code issued by the National Pension Commission (PENCOM) in 2008.
- The Code of Corporate Governance for the Insurance Industry issued by the National Insurance Commission (NAICOM) in 2009.

The SEC constituted a National Committee in 2008 and mandated the committee to examine the SEC 2003 Code with a view to addressing its shortcomings and to further enhance corporate governance mechanisms for its enforceability. The committee was also directed to identify causes and weaknesses in the corporate governance mechanisms in Nigeria and to examine and recommend ways to achieve greater compliance and to further advise on the numerous issues that are momentous in promoting comprehensive corporate governance practices in Nigeria (Olutuyi 2017).

The committee submitted its report in 2009 to SEC, which was accompanied by a draft Revised Code of Corporate Governance. SEC, after consultations with other regulatory bodies, appraised the draft of the committee and made certain amendments at its 43rd meeting. The SEC 2003 Code was replaced with the Code of Corporate Governance in Nigeria on the 1st of April, 2011 by the SEC. The SEC 2011 Code focused on five key principles, namely, leadership, effectiveness, accountability, remuneration, and relations with shareholders. The dynamic nature of the capital market and corporate world challenges compelled the SEC to amend the 2011 code further to reflect the international best practices which came into effect on May 12, 2014, as the SEC Code of Corporate Governance for Public Companies (Olutuyi 2017).

A few months after the 2011s Code became operational, the Financial Reporting Council of Nigeria Act 2011 (FRC Act), was enacted by the Federal government. The FRC Act contained comprehensive provisions regarding the operation of companies in Nigeria and provided for the establishment of the Financial Reporting Council of Nigeria responsible for issuing, monitoring, and enforcing accounting, auditing, actuarial, and governance standards concerning Nigerian public companies and public interest entities. The council tried to address the issue of multiplicity of overlapping legislation and unified the sectoral corporate governance codes with the National Code of Corporate Governance 2016 (NCCG), released in October 2016. The federal government suspended the NCCG in November following firm opposition from various stakeholders (Olutuyi 2017).

Ghana

Prior to its independence, Ghana was a British colony, and post-independence, many rules and regulations in Ghana were based on the UK legislation. The Ghanaian Companies Code, 1963 (Act 179) which specifies the requirements of corporate governance essentials in Ghana such as greater roles for the company's directors and effective participation of the shareholders in company's affairs is based largely on the English Companies Act of 1948 (Adda and Hinson 2006). While revisions have been made to the Code in 2006 and in 2012, there have been no major changes since the enactment.

The Ghana Securities and Exchange Commission issued its Corporate Governance Guidelines on Best Practices (SEC Code) in 2002. It was developed in reference to the code of best practices issued by regulatory bodies and applies to all listed corporate bodies and companies licensed under the Securities Industry Law. The Ghana SEC (2002) identified some common elements that underlie good corporate governance upon which further evolution and developments in governance structures are built today. They include the rights of shareholders, the equitable treatment of shareholders, the roles of stakeholders, disclosure, transparency, and the responsibilities of the board (Adegbite 2012).

The Bank of Ghana (BoG) also incorporated some corporate governance rules into the Banking Act, 2004 (Banking Act, 2004 was amended in 2007) and published a Code of Conduct for Primary Dealers in Government Securities in 2011 (ACGN 2016). The Ghanaian regulatory framework about corporate governance comprises of the following (Tsamenyi et al. 2007):

- Ghana's companies code (1963) revised in 2006 and 2012.
- Securities Industry Law, 1993 as amended by the Securities Industry Act (2001) (Act 590), and SEC Regulations 2003, which governs issuers and securities industry in Ghana.
- The Membership and Listing Regulations of the Ghana Stock Exchange (GSE 1990).

The Ghanaian regulatory framework is also supported by the Ghana national accounting standards and the codes of professional conduct imposed by the Institute of Chartered Accountants (Ghana) on its member (Tsamenyi et al. 2007).

The Ghana Public Services Commission (PSC) published the Corporate Governance Manual for Governing Boards/Councils of the Ghana Public Services in 2015, which addresses corporate governance for Public Services organizations (PSC, ACGN 2016).

2.5 Concluding Thoughts

Governance practices have evolved over the ages and have also undergone a significant transformation. Corporate governance has been in the picture for a few centuries with the advent of corporations. However, it has evolved with the evolution of corporations and in the face of transgressions, financial scandals and economic disasters, and along with it has also evolved its subset disciplines: operational governance, financial governance, HR governance, security governance, risk governance, IT governance, and data governance.

Finances, human resources, operations, security and risks have been in existence from the beginning of civilization, and governance can be traced back to the beginning of mankind and civilization, with governance taking the form of the ruling class and monarchy.

However, what is relatively new is the governance of information technology and electronic data. If data is equated with facts, then data has been there since the beginning of mankind but on a much smaller scale and in a different form. Therefore, data has not been a candidate for active governance until the last few decades.

In the next chapter we discuss these different subdisciplines of corporate governance, the concept of data and data classification, the evolution of data governance and how data governance ties to other corporate governance subdisciplines.

References

ACGN (2016) State of corporate governance in Africa: an overview of 13 countries. https://www.afcgn.org/wp-content/uploads/2016/03/ACGN-Corporate-Governance-Report-Feb-2016.pdf. Accessed 4th July 2018

Adda E, Hinson R (2006) Exploring the legal framework for business and ethical practices in Ghana. J Leg Ethical Regul Issues 9(2)

Adegbite OG (2012) Corporate governance developments in Ghana: the past, the present and the future, vol 1(2). Public and Municipal Finance, pp 71–75. https://businessperspectives.org/images/pdf/applications/publishing/templates/article/assets/4290/pmf_2012_02_Adegbite.pdf. Accessed 3 June 2018

Altintaş NN (2010) Evolution of corporate governance in the United States of America, Istanbul University. J Soc Sci 153–161

Ancient India Government, Cultural India. https://www.culturalindia.net/indian-history/ancient-india/ancient-government.html Accessed 28 May 2018

Arsalidou D, Wang M (2005) Difficulties with Enforcing Western Standards of Corporate Governance in Asia. Euro Bus Law Rev 2:329–340

Arthashastra, Book I, Chapter XIX, p 39

ASX (2003) Principles of good corporate governance and best practice recommendations. https://www.asx.com.au/about/pdf/ASXRecommendations.pdf

ASX Corporate Governance Council (2003) Principles of good corporate governance and best practice recommendations. Australian Stock Exchange Ltd, Sydney

BG Attorneys (2018) Bowman gilfillan attorneys. Quick guide to corporate governance and king III. https://services.bowman.co.za/Brochures/OnlineServices/CorporateGovernance/Corporate-Governance-King-3.pdf. Accessed 1 July 2018

Baily MN, Klein A, Schardin J (2017) The impact of the dodd-frank act on financial stability and economic growth. Russell Sage Found J Soc Sci 3(1):20

Balkhi S (2014) 25 biggest corporate scandals ever, updated on July 15, 2014. https://list25.com/25-biggest-corporate-scandals-ever/5/. Accessed 9th May 2018

Bank For International Settlements (2005) Enhancing corporate governance for banking organizations, ISBN print: 92–9131–699–7. https://www.bis.org/publ/bcbs122.pdf. Accessed 28 Apr 2018

Becht M, Bolton P, Röell A (2004) Corporate governance and control (October 2002; updated August 2004). ECGI—Finance Working Paper No 02/2002

Benes N (2013) The roles and functions of Kansayaku boards compared to audit committees. ACGA Paper on Kansayaku and Audit Committees. Asian Corporate Governance Association (ACGA), Hong Kong

Berle A, Means GC (1932) The modern corporation and private property, 1st edn. Brace and World, Harcourt

Borak D (2017) House votes to kill Dodd-Frank. Now what?, 8 June 2017. Archived from the original on 29 Nov 2017. Accessed 9 June 2017. …advanced the 'crown jewel' of the GOP-led regulatory reform effort, effectively gutting the Dodd-Frank financial regulations that were put in place during the Obama administration

Borak D, Barrett T (2018) Senate votes to roll back parts of Dodd-Frank banking law. cnn.com. Archived from the original on 23 Mar 2018. Accessed 8 May 2018

Brunelli M, Poor reference data management causing headaches in financial sector, TechTarget, E-guide, Bloomberg Vault. https://docs.media.bitpipe.com/io_10x/io_108208/item_640879/Bloomberg_sDataManagement_IO%23108208_E-Guide_021813.pdf

Bryan B (2017) The House quietly voted to destroy post-financial-crisis Wall Street regulations, 9 June 2017. Business Insider. Archived from the original on 29 Nov 2017. Accessed 9 June 2017

Bosch H (1991) Corporate practices and conduct. Information Australia, Melbourne

Cadbury A (1992) Report of the committee on the financial aspects of corporate governance. Gee Publishing, London

Chandragupta Maurya Wiki. https://en.wikipedia.org/wiki/Chandragupta_Maurya

Cheffins B (2011) The history of corporate governance. SSRN Electron J. https://doi.org/10.2139/ssrn.1975404

Cheffins BR (2013) The history of corporate governance. In: Wright M, Siegel D, Keasey K, Filatotchev I (eds), The Oxford handbook of corporate governance (OUP 2013)

Cheffins BR (2015) The rise of corporate governance in the UK: when and why , draft version. https://www.repository.cam.ac.uk/bitstream/handle/1810/253481/Cheffins%202015%20Current%20Legal%20Problems.pdf?sequence=1&isAllowed=y. Accessed 31 May 2019

Chen J (2015) A Primer on Corporate Governance. Business Expert Press. ISBN: 9781631572296

China Confederation of Enterprises (1999) The comments and proposals from SOE managers. Zhongguo Jingji Shibao [China Economic Times], Sept 2

Colpana AM, Yoshikawab T, Hikinoc T, Miyoshid H (2007) Japanese corporate governance: structural change and financial performance. Asian Bus Manage Suppl 1(6):S89–S113. https://doi.org/10.1057/palgrave.abm.9200240

consumerfinance.gov prepared opening statement of CFPB director Richard Cordray before the house committee on financial services—Consumer Financial Protection Bureau. Archived from the original on 4 May 2018. Accessed 8 May 2018

Cumming D, Johan S (2018) The Oxford handbook of IPOs. Oxford University Press, 13 Dec 2018

Curry T, Shibut L (1999) The cost of the savings and loan crisis: truth and consequences. FDIC Banking Review

Darrough MN (2004) The FCPA and the OECD convention: some lessons from the US. Experience. https://papers.ssrn.com/sol3/papers.cfm?abstract_id=555643 Accessed 8 May 2018

De Jong A, Gelderblom O, Jonker J (2010) An admiralty for Asia: Isaac le Maire and conflicting conceptions about the corporate governance of the VOC (Working Paper Erasmus Research Institute of Management)

Dignam (1998b) Co Law 19:140(141)
Dignam (1998a) Co Law 19:140
Du Plessis JJ, McConvill J, Bagaric M (2005) Principles of contemporary corporate governance. Cambridge University Press, New York. ISBN 0521617839
Eells RSF (1960) The meaning of modern business: an introduction to the philosophy of large corporate enterprise. Columbia University Press, NY
Eisenberg M (1976) The structure of the corporation: a legal analysis. Little Brown, Boston
European Commission (2003) Modernizing company law and enhancing corporate governance in the European Union—a plan to move forward, COM (284)
Fernando AC (2011) Corporate governance: principles, policies and practices, 2nd ed. Pearson India
Financial Services Authority (FSA) (2009) The Turner Review a regulatory response to the global banking crisis, Mar 2009. https://webarchive.nationalarchives.gov.uk/20090320232953/http://www.fsa.gov.uk/pubs/other/turner_review.pdf Accessed 8 May 2018
Financial Times (2017) Did Dodd-Frank really hurt the US economy? Financial times, 13 Feb 2017. Archived from the original on 24 Jan 2018. Accessed 23 Jan 2018
Fleming G (2003) Corporate governance in Australia. Agenda 10(3):195–212
Frentrop P (2003) A history of corporate governance, 1602–2002. Deminor, Brussels
Funnell W, Robertson J (2013) Accounting by the first public company: the pursuit of supremacy, Routledge. ISBN:0415716179
Gakeri JK (2013) Enhancing Kenya's securities markets through corporate governance: challenges and opportunities. Int J Hum Soc Sci 3(6):96–117
Gallegos F (2003) IT audit independence: what does it mean? Information systems audit and control association. www.isaca.org, https://pdfs.semanticscholar.org/7977/ce8afa66d4f248b845ce40c612745069c3eb.pdf?_ga=2.232170975.1906607567.1582434687-95003158.1580095176 Accessed 23 Feb 2020
GAO Report (1981) U.S. GEN. ACCOUNTING OFFICE, AFMD-81-34, Impact of Foreign Corrupt Practices Act on U.S. Businesses, https://www.gao.gov/assets/afmd-81-34.pdf Accessed 10 May 2019
Gelderblom O, De Jong A, Jonker J (2010) Putting Le Maire into perspective: business organization and the evolution of corporate governance in the dutch republic, 1590–1610. In: Koppell J (ed) Origins of shareholder advocacy. Palgrave Macmillan, New York
Ghana SEC (2002) Code of best practices for the public companies
Greenbury R (1995) Report on directors' remuneration. Gee Publishing, London
Hermes EOS (2017) Corporate governance principles China Hong Kong. https://www.hermes-investment.com/wp-content/uploads/2018/10/china-cg-principles-2017-dec-2017.pdf Accessed 8 Dec 2019
Holt AL (2013) Governance of IT—an executive guide to ISO/IEC 38500, BCS Learning & Development Limited. ISBN: 9781780171548
Holton GA (2002) History of value-at-risk: 1922–1998, working paper. https://stat.wharton.upenn.edu/~steele/Courses/434/434Context/RiskManagement/VaRHistlory.pdf Accessed 10 May 2019
Institute of Directors in Southern Africa Guide to questions and answers on King IV (PDF). https://c.ymcdn.com/sites/www.iodsa.co.za/resource/collection/16F4503D-86F9-43D5-AEB4-C067B06EB59C/Guide_to_questions_and_answers_on_King_IV.pdf Accessed 7 Mar 2018
Institute of Directors of Southern Africa (2016) The King IV report on corporate governance for South Africa 2016. https://c.ymcdn.com/sites/www.iodsa.co.za/resource/resmgr/king_iv/King_IV_Report/IoDSA_King_IV_Report-WebVe.pdf Accessed 3 Mar 2018
IOG. https://iog.ca/defining-governance/
Itami H (2005) Revision of the commercial code and reform of the Japanese corporate governance. Jpn Labor Rev Tokyo 2:4–25
Jason (2018) Biggest Corporate Scandals Ever, October 25, 2018. https://list25.com/25-biggest-corporate-scandals-ever/. Accessed 23 Feb 2020

Jones G, Zeitlin J (2008) The Oxford handbook of business history, 25 Jan 2008. Oxford University Press

Kaushik CNR Corporate governance—its origin in ancient India. DRIEMS Business Review, Vol 1 No 1. https://www.driems.ac.in/mba/Download/6%20CORPORATE%20GOVERNANCE%20-%20ITS%20ORIGIN%20IN%20ANCIENT%20INDIA.pdf Accessed 7th May 2018

KCGPC (2015) Kenya's Corporate Governance Practices Code. https://www.bowmanslaw.com/wp-content/uploads/2016/08/Kenyas-Corporate-Governance-Practices-Code-2015.pdf Accessed 1 July 2018

Kerandi A (2008) Governance agenda for sub-saharan Africa issues and challenges. Fed Govern 5:1

Keudell F von (2012) The internet: the book of revelations, vol 5

Kiel G, Nicholson G, Beck J, Tunny JA (2012) Directors at work: a practical guide for boards. Publisher, Thomson Reuters

Kingston City Group, Corporate governance developments in the UK. https://www.kingstoncitygroup.co.uk/includes/docs/Corporate%20governance%20developments%20in%20the%20UK%20-%20Overview.doc Accessed 1 May 2019

Koh J, Yip A (2015) Evolution of corporate governance in Singapore, Board Room Matters, vol 2. Singapore Institute of Directors, Singapore. https://www.eguide.sid.org.sg/images/SIDeGuide/articles/BM/1.%20Boardroom%20Matters%20Vol%20II-The%20Evolution%20Of%20Corporate%20Governance%20In%20Singapore.pdf Accessed 1 May 2020

KPMG (2016) Corporate governance in Japan 2016. https://home.kpmg.com/content/dam/kpmg/jp/pdf/jp-en-corporate-governance-overview-2016.pdf Accessed 26th June 2016

Kuschnik B (2008) The Sarbanes Oxley act: big brother is watching you or adequate measures of corporate governance regulation? 5 Rutgers Bus Law J 64–95. https://businesslaw.newark.rutgers.edu/RBLJ_vol5_no1_kuschnik.pdf Accessed 1 July 2018

Landler M (2003) Scandal outrages Europeans; solutions may be patchwork. The New York Times, https://www.nytimes.com/2003/12/25/business/scandal-outrages-europeans-solutions-may-be-patchwork.html

Lannoo K, Khachaturyan A (2003) Reform of corporate governance in the EU, CEPS POLICY BRIEF NO 38. https://aei.pitt.edu/1974/1/PB38.pdf Accessed 10 July 2020

Lannoo K, Khachaturyan A (2004) Reform of corporate governance in the EU, European Business Organization Law Review (EBOR), vol 5, No 1, p 37

Library of Congress (2009) Bill summary & status—111th congress (2009–2010)—HR4173—all information—THOMAS (Library of Congress). Archived from the original on Aug 19, 2014. Accessed 22 July 2010

Nottage L, Wolf L, Anderson K (eds) (2008) Corporate Governance in the 21st century: Japan's gradual transformation (Edward Elgar, Cheltenham et al, forthcoming Sept 2008). https://eprints.qut.edu.au/90105/1/elgar_book_intro.pdf

Lukomnik J (2016) Thoughts on the origins and development of the modern corporate governance movement and shareholder activism (chapter 22, pp 450–460). In: Leblanc R (ed) The handbook of board governance: a comprehensive guide for public, private and not-for-profit board members. Wiley

Marshall L, Greene R (2015) The state and fate of community banking. Harvard Mossavar-Rahmani center for business and government. Archived from the original on 20 Jan 2017. Accessed 21 Jan 2017

Mallin C (2006) International corporative governance: a case study approach. Edward Eldar Publishing Ltd., Northampton, Massachusetts, USA

New York Times (1977) Management: the first draft of a new corporate constitution. New York Times, 22 Apr 1977, 91

Marshall JB (2015) Corporate governance practices: an overview of the evolution of corporate governance codes in Nigeria. Int J Bus Law Res 3(3):49–65. Seahi Publications

Martin JG (2004) Compliance with the foreign corrupt practices Act of 1977 in the Post-Sarbanes Oxley world. https://www.winstead.com/articles/articles

Mees B (2010) Corporate governance as a movement. Paper presented at the second annual conference of AAHANZBS, The University of Sydney

Mees B, Ramsay IM (2008) Corporate regulators in Australia (1961–2000): From companies' registrars to ASIC. Australian J Corp Law 22(3):212–254

Mishra SN (2003) Public governance and decentralization, vol 1. Mittal Publications. p 935. ISBN 81-7099-918-9

Morikawa H (1992) Zaibatsu, the rise and fall of family enterprise groups in Japan. University of Press Tokyo, Tokyo

Mueller DC (2012) (ed) The Oxford handbook of capitalism. Oxford University Press, p 333. ISBN:0195391179

Mukabwa J (2016) Corporate Governance in Kenya (Part I). https://www.linkedin.com/pulse/corporate-governance-kenya-part-i-james-mukabwa/. Accessed 7 June 2018

Muniapan B, Dass M (2008) Corporate social responsibility: a philosophical approach from an ancient Indian perspective. Int J Indian Cult Bus Manag 1(4):408–420

Nader R, Green M, Seligman J (1976) Taming the giant corporation. Norton and Company Inc., W.W

Naidoo R (2002) Essentials of corporate governance for South African Companies. Double Storey, Cape Town

Naspers Application of and approach to King III. Archived from the original on 23 Aug 2011. Accessed 4 Aug 2011

Nat OFO (2013) Historical development of corporate governance in Nigeria' (2013). The Corporateprof.com/historical-development-of-corporate-governance-in-nigeria. Accessed 22 Feb 2018

Nbcnews.com US Senate approves bill rewriting post-crisis bank rules. Archived from the original on 15 Mar 2018. Accessed 8 May 2018

Nikko Asset Management (Nikko am), Where are we with corporate governance in Japan? https://sustainability.nikkoam.com/list-esg-news/where-are-we-with-corporate-governance-in-japan Accessed 1 Dec 1 2019

O'Conor M (2005) Network security, the implications of Sarbanes Oxley for IT professionals in the UK

Ocasio W, Joseph J (2005) Cultural adaptation and institutional change: the evolution of vocabularies of corporate governance, 1972–2003. Poetics 33:163–178

OECD (2004) OECD principles of corporate governance

Olutuyi O (2017) How corporate governance can affect Nigeria's development, 27 July 2017. Financial Nigeria Blog. https://www.financialnigeria.com/how-corporate-governance-can-affect-nigeria-s-development-blog-273.html. Accessed 2nd July 2018

Opinion Front, 14 biggest accounting scandals of all time that you cannot ignore. https://opinionfront.com/biggest-accounting-scandals-of-all-time Accessed 23 Feb 2020

Ovsiannikov K (2017) Corporate governance reforms in Japan: instilling the New Regime. Cogent Bus Manage 4(1)

Parker S, Peters G, Turetsky H (2002) Corporate governance and corporate failure: a survival analysis. Corp Gov 2:4–12. https://doi.org/10.1108/14720700210430298

PSC Public Services Commission, Corporate Governance Manual for Governing Boards/Councils of the Ghana Public Services. https://www.psc.gov.gh/file/Corporate%20Governance%2027.02.15.pdf. Accessed 3rd May 2018

PSICG, Private sector initiative for corporate governance, principles for corporate governance in Kenya and a sample code of best practice for corporate governance. ISBN 9966-9969-0-12. Accessed 30th July 2016

WileyRein & Fielding (2004) Foreign corrupt practices act (FCPA) handbook. Excert. https://www.wiley.law/resources/documents/fm13935.pdf

Republic of South Africa (1973) Companies Amendment Act No 61. https://www.acts.co.za/company/index.htm Accessed 05 Oct 2007

Republic of South Africa (1990) Banks Act No 94. https://www.reservebank.co.za/internet/Publication.nsf/LADV/0B2DBBD25565FE0242256D8D00341E9D/$File/Banks+Act+1990.pdf Accessed 19 sept 2008

Republic of South Africa (1998) Insider trading Act No 135. https://www.info.gov.za/view/DownloadFileAction?id=70669 Accessed 19 Sept 2008

Republic of South Africa (1999) Public Finance Management Act No 1. https://www.wrc.org.za/downloads/legislature/PFMA1-9.pdf?bcsi_scan_6E709EC7EBBC9476=1 Accessed 19 Sept 2008

Republic of South Africa (2004) Companies amendment act no 20. RSA, Cape Town

Republic of South Africa (2006) Corporate laws amendment act no 24, 2006. RSA, Cape Town

Rezaee Z (2018) Corporate Governance in the Aftermath of the Global Financial Crisis, vol IV. Business Expert Press, 4 Apr 2018

Rezaee Z, Zhang R, Saadulah S, Zeigenfuss D (2011) Corporate governance: an analysis of existing syllabi, Mar 2011. IUP J Gov Publ Policy 6(1):1–30

Rosenman S (ed) (1938) The Public Papers of Franklin D Roosevelt, volume two: the years of Crisis, 1933. Random House, New York

Business Roundtable (1978) The role and composition of directors of the large publicly owned corporation. Bus Law 33:2083, 2089, 2108

Ruparelia R, Njuguna A (2016) The Evolution of corporate governance and consequent domestication in Kenya. Int J Bus Soc Sci 7(5)

Schwartz DJ (1976) Penn central: a case study of outside director responsibility under federal securities law. UMKC Law Rev 45:394–421

Seligman J (1982) The transformation of wall street: a history of the securities and exchange commission and modern corporate finance. Houghton, Mifflin, Boston

Seligman J (1987) A sheep in wolf's clothing: the American law institute principles of corporate governance project. George Washington Law Rev 55:325–81

Shanti Parva Wiki. https://en.wikipedia.org/wiki/Shanti_Parva

Sharma SK (2003) Good Governance in Ancient India: Remembering Kingship in Shantiparavam of Mahabharat. Meerut J Polit Sci Publ Adm Meeurt, Jan-June 2003, vol VI, no-1, p 109

Small ML (2011) The 1970s: the committee on corporate laws joins the corporate governance debate. Law Contemp Probl 74:129–136

Solomon J, Solomon A (2004) Corporate governance and accountability. Wiley, Chichester

South African Institute of Chartered Accountants (2010) An integrated report is a new requirement for listed companies, 4 June 2010. Accessed 8 Aug 2011

Standards Australia (2003) Good governance principles (AS 8000–2003). Standards Australia International Ltd., Sydney, p 6

Standard and Poor's (2002) Corporate Governance In Hong Kong?, 22 Jan 2002. https://www.legco.gov.hk/yr01-02/english/panels/fa/papers/fa0724cb1-2513-1e.pdf Accessed 1 Dec 2019

Stewart Kyd (1794) A Treatise on the Law of Corporations, p 13

Subcommittee on the Functions and Responsibilities of Directors (1976) Committee on corporate laws, section of corporation, banking and business law. Corporate directors' guidebook. Bus Lawyer 32:5–52

Suda JP (1970) Dharma: its nature and role in ancient India. Indian J Polit Sci 356–366

Takahashi E, Shimizu M (2005) The future of japanese corporate governance: the 2005 reform. J Japan Law 19:35

Tarantino A (2008) Governance, risk, and compliance handbook: technology, finance, environmental, and international guidance and best practices. Wiley, ISBN 9780470095898

Tarmy S (2017) Data governance in financial services is make-or-break for compliance, 15 Sept 2017. IO-Tahoe Blog. https://io-tahoe.com/financial-services-compliance/

Tarun RW (2006) Basics of the foreign corrupt practices act. Latham & Watkins, Chicago

TCFACG-GEE (1992) Committee on the financial aspects of corporate governance. Sir Adrian Cadbury, chair. Report. Gee, London

The Institute of Directors of Southern Africa (1994) King report on corporate governance for South Africa. IOD, Johannesburg

The Institute of Directors of Southern Africa (2002) King report on corporate governance for South Africa. ISBN 0-620-2885

The Institute of Directors of Southern Africa (2009) King Report on Governance for South Africa—2009

Thompson L (2001) A history of South Africa, 3rd edn. Yale University Press, USA

Treasury HM (2009) The Walker review of corporate governance in UK banks and other financial industry entities

Tricker RI (1984) Corporate governance: practices, procedures and powers in British Companies. Gower Publications

Tsamenyi et al (2007) Disclosure and corporate governance in developing countries: evidence from Ghana. Manag Audit J 22(3):319–334

TSE (2004) Principles of corporate governance for listed companies. Tokyo Stock Exchange, Inc. https://www.jpx.co.jp/english/equities/listing/cg/tvdivq0000008j6d-att/principles.pdf

Turner L, Weickgenannt A (2013) Accounting information systems: the processes and controls, 2nd edn. Wiley

Ville S, Merrett D (2000) The development of large scale enterprise in Australia, 1910–1964. Bus Hist 42:13–46

Vinten G (1998) Corporate governance: a UK/USA comparison 19(6). Credit Control 18:18–19

Wiki1. https://en.wikipedia.org/wiki/Corporate_governance_of_information_technology

Wiki2. Dodd frank wall street reform and consumer protection act. https://en.wikipedia.org/wiki/Dodd%E2%80%93Frank_Wall_Street_Reform_and_Consumer_Protection_Act

Wiki3. https://en.wikipedia.org/wiki/Turnbull_Report

Wiki-Hampel, Hampel Report. https://en.wikipedia.org/wiki/Hampel_Report

Wiki-King Report. https://en.wikipedia.org/wiki/King_Report_on_Corporate_Governance

Wiki-UK-CGC. UK Corporate Governance Code, https://en.wikipedia.org/wiki/UK_Corporate_Governance_Code

World Bank (1989) Sub-Saharan Africa: from crisis to sustainable development. World Bank, Washington

Wright M, Siegel D, Keasey K, Filatotchev I (2013) The Oxford handbook of corporate governance. OUP Oxford

Yahoo.com (2018) Senate passes bill easing Dodd-Frank rules for banks. Archived from the original on 16 Mar 2018. Accessed 8 May 2018

Yao Y (2009) Historical dynamics of the development of the corporate governance in Japan. J Polit Law 2(4)

Yousuf S, Islam MA (2015) The concept of corporate governance and its evolution in Asia. Res J Finc Account 6(5). ISSN 2222–1697 (Paper) ISSN 2222–2847 (Online). https://www.iiste.org/Journals/index.php/RJFA/article/viewFile/20768/21359

Yu B, Rudge L, Grahma D (2014) Hong Kong Corporate Governance: a practical guide. Whitepage Publications. ISBN9780956584250; https://www.slaughterandmay.com/media/2170661/hong-kong-corporate-governance-a-practical-guide.pdf

Chapter 3
Corporate Governance Subdisciplines, Data, and Data Governance

> A steer may not accelerate the speed of vehicle directly; by navigating the better path, it can shorten the distance or save the energy to reach the right destination.
> —Pearl Zhu

Abstract Corporate governance is a broad discipline that has several subdisciplines which include, but are not limited to operational governance, financial governance, HR governance, risk governance, security governance, risk governance, IT governance, and data governance. This chapter discusses the corporate governance approach and the different subdiscplines of corporate governance. This chapter also introduces the readers to the concepts of data and data classification. The evolution of data governance, the lack of standardization of data governance definition, and the difference between IT governance and data governance are also discussed. A separate section explores how data governance ties to other corporate governance subdisciplines.

3.1 Corporate Governance Subdisciplines—Introduction

Corporate governance is the system of policies, rules, mechanisms, practices, authorities, and processes used to direct, manage, and control a corporation. The objective of corporate governance is to enable an effective and prudential management to deliver long term corporate and financial success. This is resonated in Toba Beta's statement—"Good corporate governance, it's about being proper and prosper."

Corporate governance became increasingly prevalent in the late 20th century and early 21st century as a result of scandals, company collapses due to unethical practices, and exposures of accounting discrepancies and fraudulent practices. The evolution of corporate governance in various geographies across the world was discussed in Chap. 2.

R. Mahanti, *Data Governance and Compliance*,
https://doi.org/10.1007/978-981-33-6877-4_3

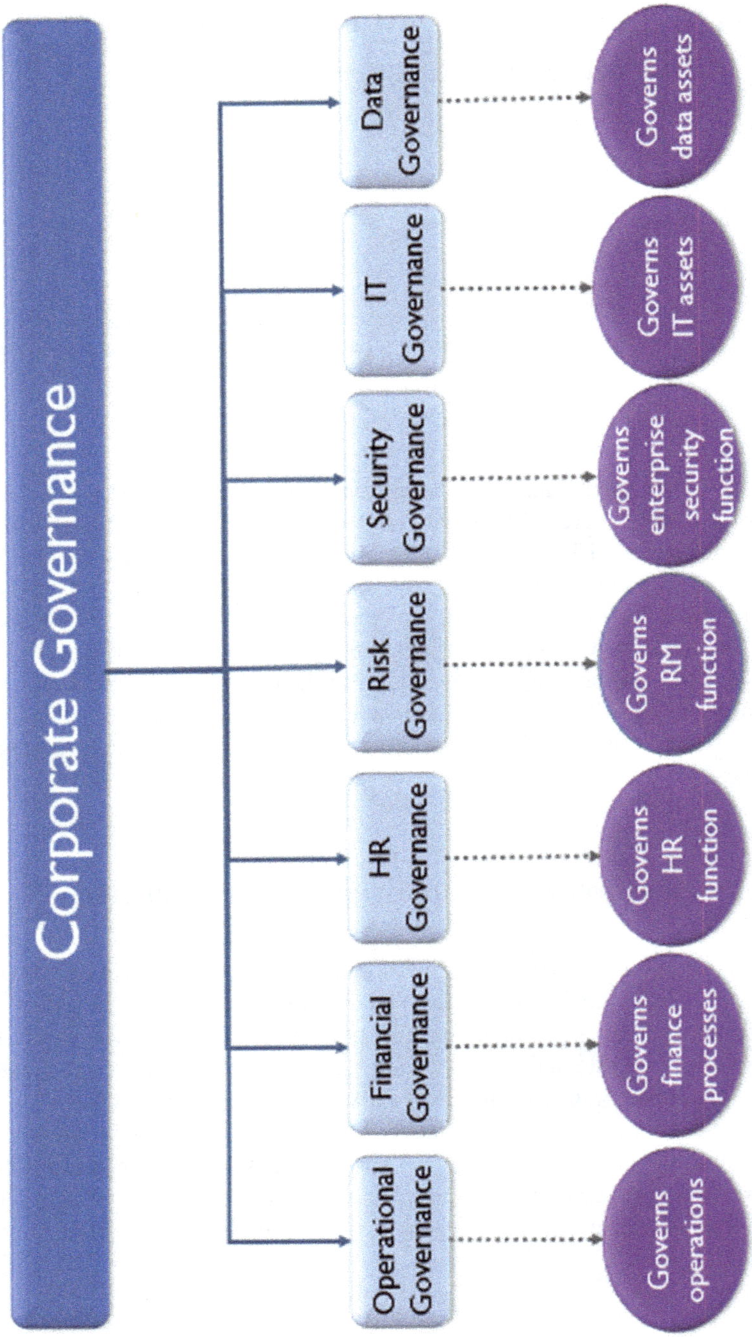

Fig. 3.1 Subset disciplines of corporate governance

Corporate governance is a broad discipline that has several subdisciplines which include, but are not limited to operational governance, financial governance, HR governance, risk governance, security governance, risk governance, IT governance, and data governance as shown in Fig. 3.1.

This chapter discusses the corporate governance approach and the different subdiscplines of corporate governance. This chapter also introduces the reader to the concepts of data and data classification. The evolution of data governance and the lack of standardization of data governance definition and the difference between IT governance and data governance is also discussed. A separate section explores data governance ties to other governance disciplines.

3.2 Corporate Governance—The Approach

There is no "one-size-fits-all" approach to corporate governance. The governance structures and framework would depend on the basic alignment of the company and the role that the company plays in the social order and the overall context. However, at a structural level, a framework for corporate governance should include the following (Barett 2001):

- Strategic direction,
- Policies and procedures,
- Control and accountability systems,
- Performance management, and
- Risk management.

The corporate governance framework should play a vital role in supporting the business and help deliver the corporate strategy. It should provide the structure through which the strategy and business objectives are set, and performance is monitored. It should have robust systems of risk management and assurance through which the risks the business faces are managed. It should include a clear framework for delegation, decision making, and accountability across the organization. It should guide the standards of behavior expected of each other (that reflect the company values and beliefs) in delivering the strategy and achieving the strategic business goals, as well as open, clear and timely communication to the shareholders. A robust corporate governance framework is indispensable to maintain good oversight and control over financial and management reporting, compliance and regulatory matters, risk management, and the approval of substantial decisions (such as acquisitions, disposals, or material agreements).

Corporate governance is the uppermost level of governance in an organization, and has various subset governance categories, which include but are not limited to the following categories-

- Operational governance,
- Financial governance,

- HR governance,
- Risk governance,
- Security governance,
- IT governance, and
- Data governance.

As a result of corporate governance efforts to better govern the leverage of corporate resources, specific attention was given to the role of information and the underpinning technology to support good corporate governance. It was soon recognized that information technology was not only an enabler of corporate governance, but as a resource, it was also a value creator that needed a better governance structure (Wiki1).

Good corporate governance helps to prevent corporate scandals, scams, fraud, and potential civil and criminal liability of the organization. It improves the reputation of the organization; and protects the interests of its stakeholders (employees, shareholders, and investors) who rely on them (Lipman and Lipman 2006). It identifies and reduces risks, reduces the opportunity for breaches, fraud, and corruption, endorses responsible decision making, ensures effective communication about strategic plans, directions and decisions, and company performance to stakeholders, and helps to ensure the protection and good management of the corporate assets.

Excellence in corporate governance is essential for the long-term performance and sustainability of organizations, and to protect and enhance the interests of the organizations' shareholders and other stakeholders.

Governance arrangements should be reviewed regularly, to reflect developments in market and industry practice, expectations, and regulations as appropriate. Standards, policies, procedures, and rules need to be responsive to the fluctuating environment and will continue to develop and evolve.

Good corporate governance can be performed cost-effectively by focusing efforts on the substantial risks faced by the organization rather than attempting to cater for any possible hypothetical risk, and by implementing and following best cost-effective practices within the organization. While there is still a huge cost involved in implementing corporate governance, the cost of not implementing the same is often higher. This is because good corporate governance results in avoidance of governmental investigations, lawsuits, fines, penalties, customer dissatisfaction, and damage to the reputation of the organization, and these benefits significantly outshine the cost of good corporate governance (Lipman and Lipman 2006).

3.2.1 Operational Governance

The economic crisis has not only been caused by financial failures but operational failures too. Thus, operational processes need to be more transparent and adequately governed to prevent operational failures, and at the same time, create value and improve operational performance by lowering operating costs, increasing efficiency, and reducing unnecessary conflicts. Operational governance is about collectively agreeing on how operating decisions will be made, executed, and monitored, and by whom. Effective operational governance results in:

- operational problems being solved in a structured manner,
- effective and timely decision making, and execution,
- better flow of information,
- reduced conflicts.

With operational governance, policies are aligned from the business to the operational aspects of the enterprise, by stating policies as business processes. This ensures that established business policies are followed throughout all organizational operations, thus enabling more effective control. Policies instituted at executive and strategic levels in the organizations (for example, by the Chief Operating Officer and business unit heads), must be consistent with policies embedded and executed in the operational infrastructure (Vitria 2008; Vagadia 2014).

3.2.2 Financial Governance

Financial governance is the rigorous set of rules, guidance, and policies issued by regulators to ensure that financial processes are well-governed. Good financial governance ensures financial sustainability and formalizes sound decision-making processes around financial management. Without appropriate financial governance, an organization is susceptible to internal and external problems, such as fraud. A large part of financial governance is a company's reporting capabilities, which include everything from internal controls, to audits, process workflows, financial controls, data tracking, security, reporting integrity, and a demonstrated consideration of critical financial governance matters can provide confidence to all stakeholders (Tagetik).

At the core of financial governance is corporate governance which emerged out of the financial crisis at different points in history like the great depression in the 1930s, the financial crisis in the mid-1980s, the global financial crisis in the late 2000s, and the regulatory measures which evolved such as the US's Sarbanes Oxley Act, UK's Companies Act, FINREP and COREP.

The best ways to ensure sound financial governance is to:

- Monitor financial performance,
- Empower audit committees, clearly defining their roles and responsibilities,
- Keep up to date and comply with all external regulatory requirements,
- Ensure automated controls are in place,
- Organize periodic risk assessments, and
- Conduct periodic internal and external audits.

3.2.3 Human Resource Governance

Human Resource (HR) governance is the human resource function dimension of the corporate governance framework. HR resources describe both the set of individuals who make up the workforce of an organization, as well as the department in the organization, which is entrusted with the responsibility for managing employee-related resources and oversees the wellbeing of the employees. In the 1960s, concepts such as motivation, organizational behavior, and selection assessments began to take shape, and the importance of labor relations began to garner a great deal of attention, which led to the creation of the term *human resources*.

The term “HR governance” was conceived much later, possibly in the mid-’90 s, along with HR’s extensive efforts tc transform it from an administrator function into a business partner function. For HR executives, Sarbanes-Oxley has many implications on HR, such as personal, legal accountability for the reliability of reporting, decision making for benefits, plans and programs, and needs HR executives to scrutinize functional operating models with the object of improving business involvement (Mercer 2003).

HR governance involves overseeing the key sensitive areas of HR, such as compensation and benefits for executives, performance management, major rules of conduct, whistleblower procedures, and labor relations. As per Mercer, HR governance is the act of leading the HR function and managing associated investments to improve the performance of the organization’s human capital assets, fulfill fiduciary and financial responsibilities, lessen enterprise HR risk, align the function’s priorities with those of the business, and enable HR executive decision making (SABPP 2017; Mercer 2003).

The distinct but interdependent elements of HR governance are effective councils, structure, and accountability (design of the council and its relationships with involved stakeholders), philosophy and operating principles (the function’s risk tolerance, approach to delegating authority, and expected level of management autonomy at the business unit or geographic levels), core management activities (such as HR strategy development, business planning, oversight of rewards plans and programs, HR resource allocation, and HR staff development/leadership succession), and performance monitoring (metrics and framework used to assess and communicate the function’s operational effectiveness, compliance, and contribution to business success) (Mercer 2003).

The main difference between corporate governance and HR governance is that while corporate governance governs the behavior of an organization as a corporate entity, HR governance ensures that employees are in line with the key principles of governance accompanied with HR best practices. Corporate governance and HR governance are strategically related as an improvement in the employee’s performance will automatically translate into an improvement in organizational performance (SABPP 2017).

By aligning the HR governance to that of organizational activities, HR creates an interrelated relationship with the business that forms the basis of its operating model and governance structure.

HR Governance focuses on the rules that govern the main components of the corporate entity referred to as the people (SABPP 2017).

3.2.4 Risk Governance

Since risk management is essential to successfully running any business, risk governance is an important subset discipline of corporate governance to set roles and accountabilities for managing risks, having transparency within the organization that risk registers are disclosed and managed, and giving assurances that the appropriate controls are in place and performing as designed (RMIA 2015).

The global financial crisis made it clear that robust risk governance mechanisms were essential to prevent such high-profile incidents from recurring. Good corporate governance requires that risks are understood, managed, and where appropriate, communicated (OECD 2014).

Risk governance is the framework within which risk management functions in a corporation. A core requirement for high-quality risk management is effective risk governance. Risk governance applies the principles of good governance to the identification, assessment, management, and communication of risks (International Risk Governance Council). It incorporates the principles of accountability, participation, and transparency in establishing policies and structures to make and implement risk-related decisions (International Finance Corporation 2013).

Risk governance is not restricted to a traditional practice of risk assessment. Still, it goes beyond it to include multiple stakeholders involvement as well as careful consideration of the comprehensive legal, political, technology, economic, social, health and safety, operations, environment, and security contexts in which risk is evaluated and managed (Renn 2008; Wiki).

Risks can be very diverse depending on the industry sector and subsector, the product or service delivered, work environment, and social and economic conditions. For example, the healthcare industry has a risk of exposing its health workers to disease and infections. Industries like technology, professional services, and financial services where white-collar jobs are predominant, risk of injury are negligible compared to manufacturing where heavy-duty machinery is usually common. Even in the manufacturing industry, risks of injury can be very diverse, with construction and mining having a high-risk environment compared to other manufacturing industries such as the food industry.

Risk governance reflects and seeks to sustain and evolve the organization's risk culture, which affects an organization's risk appetite as well as attitudes towards risks. Good risk governance should result in risk being accepted and managed effectively within the organization's identified and agreed to risk appetites. Effective risk governance outlines the risk accountabilities, roles, and responsibilities of the organization's workforce and hence plays a fundamental role in entrenching the correct risk values and culture in an organization.

Risk governance should put in place a structure of risk responsibility for both the direct and indirect financial risks, and non-financial risks (such as conduct and compliance) throughout the organization. This ensures that every employee in the organization is cognizant of their respective risk responsibilities and accountabilities, and those of the others with whom they work (The Institute of Operational Risk). A comprehensive risk governance framework promotes a clear understanding and documentation of the risk appetite of an organization, that is the types and levels of risks that an organization can take and is willing to take to achieve its goals, and provides transparency in the manner in which the employees implement their responsibilities. Risk governance in an organization should cover all aspects of risk management, which includes establishing its risk appetite, identification, and assessment of risks, prioritization of risks based on the likelihood of occurrence and severity of impacts, risk mitigation activities, the establishment of risk policies and controls, and continuous monitoring and reporting.

For an organization to realize the benefits of effective risk governance and risk management, the Board and senior management should treat risk management as a high priority and must be strongly committed to their risk governance responsibilities. Their commitment influences the risk culture of the organization. The greater the levels of Board commitment, the healthier the organization's risk culture.

The Board and senior management should define and allocate accountabilities, roles, and responsibilities for the different risk management functions to make sure that all these functions are carried out in an effective and proficient manner. The Board should also ensure that sufficient processes are in place to ensure the organization's adherence to approved risk policies. While the Board is ultimately accountable and responsible for risk governance and answerable for any lapse of the same, effective risk governance and oversight commences with a clear common understanding of the agreed and approved risk appetite, agreement and demarcation of the risk governance responsibilities amongst the Board, senior management and other stakeholders, with the senior management being responsible for implementing the policies, procedures, and having appropriate metrics, controls, monitoring and reporting in place to achieve risk governance within the organization. Relevant good quality data can help organizations to spot potential risks early and enable preventive actions accordingly.

3.2.5 Security Governance

Security governance is the security function dimension of the corporate governance framework. It includes rules, policies, processes, roles and responsibilities, and decision rights to direct and control an organization's approach to protecting its physical and digital assets as well as protecting the organization's public reputation. Effective security governance results in effective co-ordination of security activities and effective security decision-making at all levels of an organization. It enables the flow of security information and communication of security decisions across the organization. Effective security governance requires good data to be able to make informed and rational security decisions.

Each organization will have a different approach to security governance depending on its strategic goals and priorities, its product and operations, security risk appetite, size, the complexity of the organization, availability of resources, and its specific legal and regulatory requirements. Also, not all assets need to be governed from a security perspective, with the same rigor. For example, with data security governance, sensitive data needs to be governed with more rigor than non-sensitive data.

An organization's security governance framework should not be approached in a silo, but should be approached while keeping in mind the other functions that need to be considered, such as HR, finance, risk, operations, and IT. It should fit and be integrated into the organization's wider governance framework. Concerning IT and data assets, security governance is covered in IT governance and data governance.

Good security governance alone will not achieve the security goals of an enterprise. Effective security governance, coupled with effective security management, will protect an organization's assets from being damaged or stolen.

3.2.6 IT Governance

IT governance or Information Technology governance is a subset discipline of corporate governance and became recognized as such in the early 1990s (Phillips 2012). IT governance needs to be designed, aligned, and linked to the principles of corporate governance. Owing to the omnipresence of information systems and information technology infrastructure throughout corporations, organizations and their business operations are largely, if not fully dependent on information technology (IT) to provide vital services. IT supports and sustains an organization's business functions and processes, and hence, IT plays a critical role in the success of an organization.

IT governance is conducted by the set of processes, procedures, policies, and technologies that are established to manage resources to align IT with business strategies. While governance developments have primarily been driven by the need for the transparency of enterprise risks and the protection of shareholder value, extensive use of IT and increased dependency on IT with IT being a crucial element in regulatory/policy compliance, has necessitated a special focus on IT governance. Hence, IT governance is a key component of corporate governance. However, IT governance cannot be practiced in isolation in an organization and needs to be aligned and integrated with the company's overall governance structure. The intent is to extend governance to IT and provide the leadership, organizational framework, and processes that ensure that the organization's IT sustains and meets the corporate strategies and objectives.

IT governance has become increasingly important and even more challenging with the use of ever-changing and increasingly competitive information technologies, such as SaaS, social media, mobile platforms, analytics, and cloud computing. IT governance is not a finite project but an onward journey towards continuous improvement and optimum efficiency where best practices, industry standards, and governance frameworks provide a set of tools and methodologies to establish good governance in a company at any level of maturity.

IT governance may be defined as a set of rules, practices, and processes that ensure the following:

- Effective and efficient use of IT in facilitating an organization to achieve business objectives,
- Ensure that IT performance is adequate,
- IT manages risks adequately and meets compliance regulations, and
- Maximizing the value of IT for the business.

Gartner defines IT governance as follows:

> IT governance (ITG) is defined as the processes that ensure the effective and efficient use of IT in enabling an organization to achieve its goals (Gartner).

The IT Governance Institute defines IT governance as follows:

> It is an integral part of enterprise governance and consists of the leadership and organizational structures and processes that ensure that the organization's IT sustains and extends the organization's strategies and objectives (ITGI 2003).

Thus, the overall objective of IT governance is to comprehend the issues and the strategic importance of IT, so that organizations can sustain their operations and implement the strategies required to extend their activities into the future. IT governance aims at ensuring that expectations and goals for IT are met, and IT risks are mitigated (ITGI 2003). The purpose of IT governance is described in the IT governance Institute Board briefing on IT governance, as to direct IT endeavors, and to ensure that IT performance meets the following objectives (ITGI 2001):

- For IT to be aligned with the enterprise and the assured benefits are attained,
- For IT to act as an enabler for the enterprise by taking advantage of opportunities and maximizing profits and hence delivering value,
- For IT resources to be used responsibly and hence ensuring optimal use of IT resources, and
- For IT-related risks to be managed appropriately and hence ensuring risk mitigation.

De Haes and Van Grembergen (2005) propose that IT governance can be implemented through a framework of structures, processes, and relational mechanisms, as shown in Fig. 3.2.

Some of the popular IT governance frameworks are the following:

- **CobiT®** Control Objectives for Information and Related Technology is regarded as the world's top IT governance and control framework. COBIT5 was published

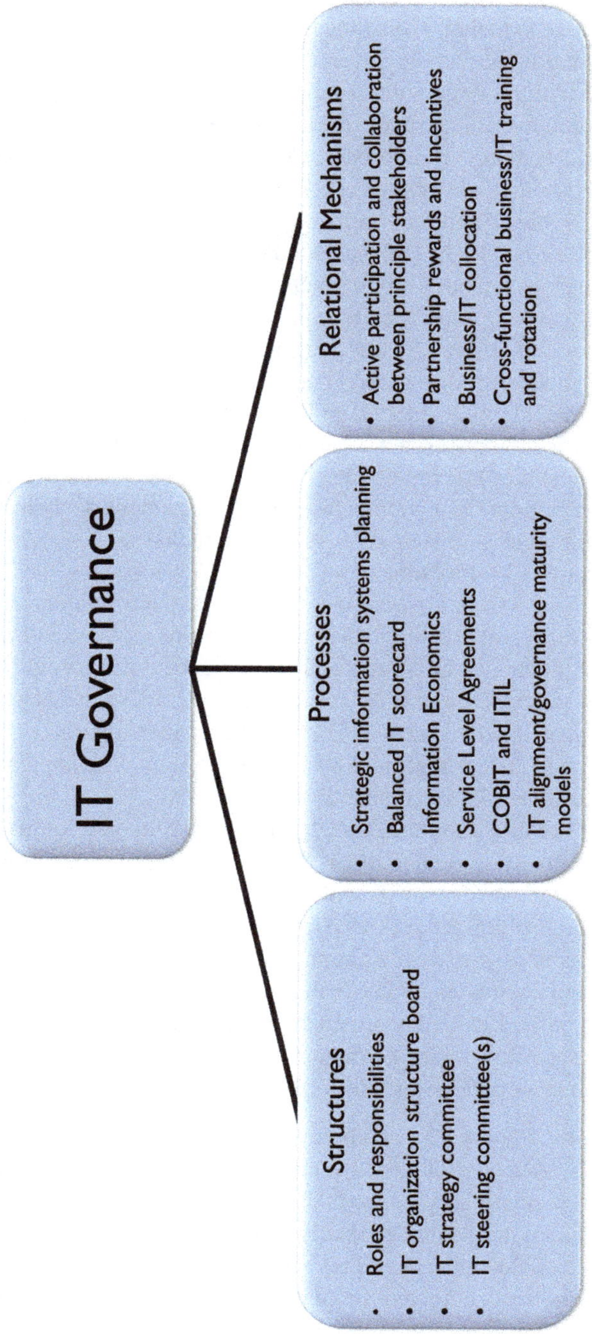

Fig. 3.2 Structures, process and relational mechanisms for implementing IT governance (adapted from De Haes and Van Grembergen 2005)

by ISACA in April 2012 as a "business framework for the governance and management of enterprise IT." It is the latest iteration of the COBiT framework and provides a reference model of 37 IT processes typically found in an organization and is split into two areas, namely governance and management. Each process is defined together with process inputs, outputs and outcomes, key process activities, process objectives, performance measures, and a maturity model to assess the capability and maturity of a process. COBIT5 consolidates COBIT4.1, Val IT and Risk IT into a single framework acting as an enterprise framework for enterprise IT governance and management. It is aligned and interoperable with TOGAF and ITIL (Wiki1). The tools included are performance measures, a list of critical success factors that outline best practices for each IT process, and maturity models for benchmarking and capability improvements (ITGI 2003).

- **ITIL®** IT Infrastructure Library®, was developed by the UK's Cabinet Office as a library of best-practice processes for IT service management. Widely adopted around the world, ITIL is supported by ISO/IEC 20000:2011, against which independent certification can be achieved by organizations (IT governance Ltd.).
- **FAIR** Factor Analysis of Information Risk (FAIR) is a comparatively new model that assists organizations in quantification and management of their risks with emphasis on cybersecurity, information, and operational risk.
- **Val IT®** Value from IT Investments (Val IT) is developed by the IT Governance Institute (ISACA). It is a governance framework of information technology-enabled business investments. It enables the generation of business value from IT-enabled investments. Val IT combines a set of practical and recognized governance principles, processes, practices, and supporting guidelines that help boards, executive management teams, and other enterprise leaders optimize the realization of value from IT investments and is designed to align with and complement COBIT. (ISACA 2008; IT Governance Institute)
- **ISO/IEC 38500** Published in June 2008, International Standard for the corporate governance of Information Technology is the *official IT governance standard* and applies to the governance of management processes and decisions involving information and communication services used by an organization (IT governance Ltd[a]). It sets out a clear and easy-to-follow framework for an organization's governance of Information and Communications Technology to align IT with organizational decisions. It can be applied to any organization, irrespective of size, sector, type, or location and regardless of their extent of IT usage, and is useful to directors, senior management as well as key staff members (IT governance Ltd[a]).

The IT governance frameworks can be used as a guide to implementing an IT governance program. Although different frameworks have been extensively implemented, there is no absolute and standard IT governance framework in place. The framework or combination of frameworks that work best for an enterprise is influenced by a multitude of elements, including business factors, organizational culture, IT maturity, as well as staff competency and expertise. The level of implementation of these frameworks also varies from organization to organization.

While it is the Chief Information Officer (CIO) who is normally accountable for implementing IT governance in an organization, the Chief Executive Officer (CEO) and board of directors must receive reports and updates to discharge their duties for IT governance and making sure that the program is operational and is providing business benefits (Smallwood 2014a).

In the past, board members overseeing IT governance was not a common practice. However, in today's world, it is a critical and mandatory responsibility (Smallwood 2014b). According to the IT Governance Institute's Board Briefing on IT governance, "IT governance is the responsibility of the board of directors and executive management," (ITGI 2003; Phillips 2012). "Board of directors" is ultimately answerable to enterprise stakeholders (ITGI 2003).

3.2.7 Data and Data Governance

Data governance is a system of policies, rules, standards, processes, practices and structures, roles and responsibilities, controls, and decision rights to oversee the management of data. Data governance should not be confused with the technical management of data.

The first English usage of the word "data" can be seen in the 1640s (OED, Wiki-Data). Technically speaking, data is the plural form of the word Latin word "datum," which means "a fact." However, in the business and IT world, "data" is most commonly used in the singular, as a mass noun. The New Oxford American Dictionary defines data first as "facts and statistics collected together for reference or analysis." Data is a collection of facts related to real-world entities, concepts, and events, represented by text, numbers, images, sound or video, and formatted in a specific way to facilitate storage, processing, retrieval, analysis, sharing, and usage.

Data has always been the king. While the concept of electronic data has been there since the advent of computer technologies, data has been around since ancient ages. The importance of data in olden days can be seen in the below quote by French Military and Political Leader, Napoléon Bonaparte,

> War is ninety percent information.

While some industry sectors are data-centric by nature and heavily rely on collecting and processing data as one of their primary activities, for example, financial services, professional services, and technology, data is equally important in industry sectors whose primary products or services do not revolve around data such as manufacturing industries as data drives or enable different processes in these industry sectors too, both before and post digitization (Mahanti 2019).

Data is an instrument of illustration. It stands for objects other than itself (Chisholm 2010). Data is both an interpretation of the objects it represents and an object that must be interpreted (Sebastian-Coleman 2013).

Data can be classified in several ways depending on its acquisition (external or internal to an organization); what it represents; uses of data; subject areas; varieties of data; criticality of data; and its sensitivity.

Master data represents the business objects that give context to business activities and transactions, but itself is non-transactional. Master data helps to answer questions like who, what, when and how, as well as expanding the ability to make sense of business activities and transactions through categorizations, groupings and hierarchies such as party, location and things (Dreibelbis et al. 2008; Wolter and Haselden 2006; Wiki-Master).

Transactional data is the data that describes the changes resulting from transactions or an event and always has a time dimension associated with it. Common examples are invoices and account transactions (deposits and withdrawals).

Reference data represents the set of permissible values that are referenced and shared by other data like master and transactional data and systems to create a standard vocabulary, structure, and format across different systems. Examples are country codes, industry codes, account status codes.

Metadata is data about data and is additional information that describes other data, like master data, transaction data, and reference data.

Data classified from a perspective of data security is the classification of data based on its level of sensitivity and the impact to the organization if the data is disclosed, altered, accessed, transported, stolen, or destroyed without authorization, as shown in Fig. 3.3. Data classification will be discussed in detail in the second book of the series: Data Governance and Data Management.

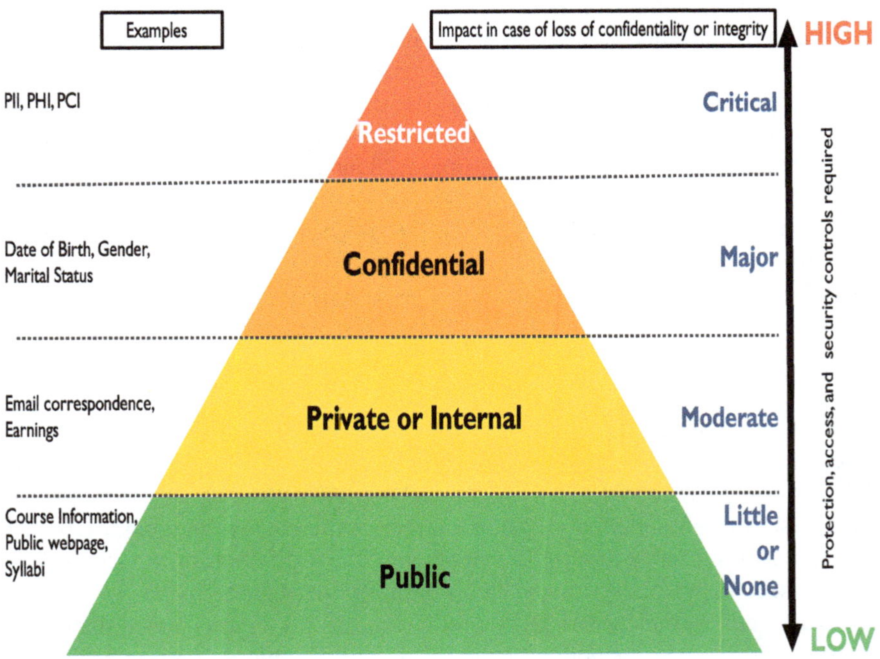

Fig. 3.3 Data Classification by data sensitivity and protection

3.3 Data Governance—History and Evolution

As data and data management have evolved over different periods of time, data governance has also evolved. In ancient times, records were maintained on leaves (for example palm leaves), before the advent of paper, with trade being the major driver in those times to track records. In the 13th and 14th centuries, the need for recording, as well as retrieving and updating data, led to the development of a double-entry book (Lake and Crowther 2013). Before the arrival of the electronic processing era, that is until later part of the 19th century, important data was manually stored in physical paper files and paper registers, and was organized in filing cabinets on a much smaller scale. There were clerks who had the express responsibility to ensure the accuracy and integrity of the data, and managers were responsible for limited governance, to ensure that the data was secured and shared as deemed proper.

The advent of electronic processing led to the electronic capture and storage of data, with punch card technology widely used at the end of the 19th century. With the advent of data management in the 1960s, systems were built to support transactional business processes like placing and processing orders, running payroll, and balancing a general accounting ledger (GL). Data was viewed as a by-product of running business and operations and had minimal value beyond the specific operation, transaction, and application that processed it. Data was not treated as a valuable and strategic enterprise asset in those times. While a few companies endeavored to govern data through enterprise data modeling, their success was restricted, as these efforts were driven by IT, without broad organizational involvement or support and authority to enforce compliance. Also, the inflexibility of packaged applications further decreased their effectiveness (Chen 2010).

While data governance is not new and has been practiced in some form as early as 1970 (Perez), the term was not popular even in the 1990s. For example, regulations have been in place even before there has been a strong awareness of the term data governance, but decisions were still needed to be made around data.

This was highlighted by Laura Sebastian-Coleman, Data Quality Lead, Finance Data Governance Center of Excellence at CVS Health when questioned about HIPAA and data governance in her statement:

> HIPAA has been in place since before there was a strong concept of "data governance." It is a good example of a challenge that required organizations to figure out how to organize decisions and solutions around data at an enterprise level.

The 1980s witnessed the evolution of technologies designed to handle data storage, data quality, and metadata management, often on a departmental or business unit basis to support different initiatives such as customer relationship management, database marketing, or data warehousing and so on. Pockets of "data governance" appeared in different parts of the organization to support these initiatives. Data governace sometimes developed as a grassroots effort, with it being occasionally popular in the IT department (SAS). As highlighted by George Firican, Director of Data Governance & Business Intelligence at UBC DAE and Founder of www.lightsondata.com, in his interview, data governance used to be considered an IT function exclusively, generally being siloed from the rest of functions/departments of the organization.

However, as the enterprise data repository continued to grow on a larger scale and customer relationship management (CRM), enterprise resource planning (ERP) system, ERP consolidations, and master data management (MDM) repositories became prevalent in the 1990s, the need for data to support multiple use cases led to conflicts. The need for data governance became evident in its role in resolving these conflicts. However, since data-related activities around this period were siloed and carried out on an application basis, data governance was siloed by individual repositories—for example, data governance for an enterprise resource planning (ERP) system, data governance for a customer relationship management (CRM) system, data governance for the data warehouse, and so on. Also, data governance was informal (Chen 2010), and was characterized by an absence of:

- distinct organizational structure,
- well-defined roles and responsibilities,
- clearly defined standards, policies, and processes, and
- metrics, or well-defined metrics.

Post the advent of the internet in the mid-1990s, transmission, and exchange of data raised new challenges around data security and data privacy. With time, as digital data varieties, volumes, and complexities continued to increase, overseeing the management of data became increasingly important. This is because while data management technologies evolved to handle the increasing volumes and varieties of data, technology, repositories, applications, and IT alone were not sufficient to oversee the management of huge volumes and varieties of increasingly complex data. A policy-process-role centric approach to data models, data quality standards, data security, and lifecycle starting from the creation/acquisition of data to the disposal of data became necessary.

The early 2000s saw data governance gaining more attention with early drivers for data governance, usually coming as directives from the compliance, risk, legal and audit departments of an organization. However, these directives were perceived by other departments in the organization as an indispensable overhead with little or no intrinsic value to the organization as a whole (SAS).

The downfall of companies like Enron, Worldcom, and Arthur Anderson led the US federal government to establish rules to improve the quality of corporate data and establishing controls and accountabilities around data. Data governance was a key component of these efforts. The accounting scandals and corporate governance failures drove the need to have better compliance, including data compliance and reporting, which in turn drove the need for better governance around the data. For example, the Sarbanes-Oxley Act required executives to be personally responsible for the data driving their businesses (SAS).

In the 2003 joint sitting of the Basel Committee on Banking Supervision (Basel), the International Organization of Securities Commissions (IOSCO), and the International Association of Insurance Supervisors (IAIS), recognized the need for better quality data which provided added impetus for implementation of data governance.

The global financial crisis drove Basel's BCBS 239 Risk Data Aggregation and Risk Reporting (RDARR) publication in 2013, which outlined the expectations of regulators from banks concerning data quality, the formal roles and responsibilities for data, and the need for standardized metadata (Pearce 2017) and controls around data.

In recent years, there has been a large increase in regulatory oversight as a result of prior corporate transgressions. Several laws and regulations have been enacted in different parts of the world, some of which are the Sarbanes-Oxley (USA), and the Dodd-Frank Wall Street Reform act (USA), the Anti-money Laundering Act (Australia), and the EU backed Basel II, Solvency II, and General Data Protection Regulation (GDPR). According to Deloitte, the number of privacy laws has grown from 20 to 100 in the past few years (i-Sight 2018). Compliance with each regulation requires high-quality data, controls, and accountabilities around data, which in turn requires effective data governance.

GDPR is the most popular example of data sovereignty regulations that are being propagated across jurisdictions, which include Australia's Privacy Act, Canada's Anti-Spam Law, the People's Republic of China's Cybersecurity Law, and Russia's Personal Data Localization law (Soares and Franco 2017). With the enforcement of the European Union's, GDPR on May 25, 2018, which focuses on the protection of personally identifiable information (PII), organizations need to have adequate data governance practices in place to be in line with its mandates on handling and securing personal data or protecting the privacy of personal data.

Multinational organizations are subject to additional regulations—not only do they have to comply with their home country regulations but also the countries in which they operate. For example, PricewaterhouseCoopers (PwC) is a global business with member firms in 157 countries. They have developed a comprehensive global data protection program and have issued an extensive "toolkit" to help the member firms comply with GDPR provisions. The member firms then implement measures applicable to their particular countries and business goals (Epler). Data governance as an information management discipline has evolved and has had to mature quickly, from being a siloed IT function to a collaborative organization-wide function between IT and business across the globe, as a result of these drivers.

The continual evolution of hardware, software, information, and communication technologies, has enabled the ability to accumulate an unprecedented amount of data. These factors along with industry trends, increased automation, and business drivers in the face of changing regulations and advent of new regulations and compliance requirements, have resulted in the paradigms for data collection, data transmission and exchange, data storage, data access and data usage to keep changing. Data governance also continues to evolve in the face of these changes.

3.4 Data Governance Definition

A crisp definition of data governance is:

> the exercise of decision-making and authority for data-related matters. (DataGovernance.com)

An elaborative definition of data governance is:

> a system of decision rights and accountabilities for information-related processes, executed according to agreed-upon models which describe who can take what actions with what information, and when, under what circumstances, using what methods. (DataGovernance.com)

No single standard definition exists for data governance. This is because of the following reasons:

- As stated by Tony Epler, Chief Data Strategist, PricewaterhouseCoopers in his interview, the responsibility, either designated or assumed, for data governance can reside in many different organizational functions within an organization. For instance, in one organization, it may reside with a specific data organization. In contrast, in other organizations, data governance might be the responsibility of IT, or OGC or Risk and Compliance, and so on. It might even be assumed by such specific functions as Finance or HR or Sales and Marketing, depending on the nature of the business and the type of data. Figure 3.4 shows some examples of different organizational functions where data governance may reside. Each organizational function generally has its own particular perspective about data management, and thus its own definition of data governance.
- The second reason for the lack of standardization, as per Tony Epler, is that an organizational definition of data governance can be highly dependent on the perception and use of data within the business. Heavily regulated industries such as health care, financial services, and manufacturing (examples- transport, electric power, pharmaceutical, and medicine manufacturing), are more concerned with things like compliance, risk reduction, standardization, data lineage, and

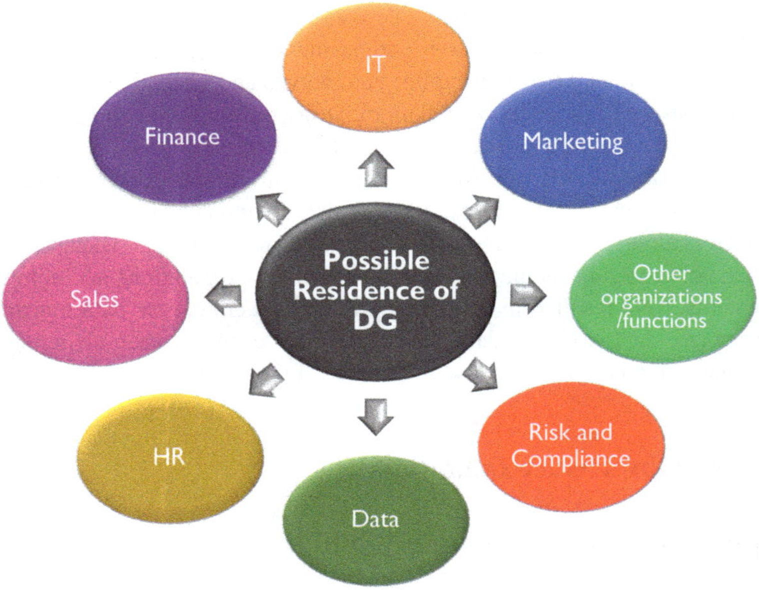

Fig. 3.4 Possible residence of Data Governance (DG)

the integrity of data flowing through the infrastructure. Less regulated industries, such as retail, must be able to react quickly to competition, so businesses are more concerned with agility, flexibility, and revenue generation than with risk and compliance. These radically different perspectives create different data strategies and processes and, thus, different ideas around data governance (Epler). In other words, business drivers, use cases, and program focus can affect the definition of data governance.

- Since data governance ties together other data management functions such as data quality management, data security management, metadata management, master and reference data management, and data integration), and other disciplines of governance such as corporate governance and sub-disciplines of corporate governance, the definition of data governance can be influenced by each of these other function and disciplines. For example, Margaret Rouse defines data governance with a data quality and security focus:

  > the overall management of the availability, usability, integrity, and security of the data employed in an enterprise. A sound data governance program includes a governing body or council, a defined set of procedures, and a plan to execute those procedures" (Rouse).

- Different perceptions, as well as an incomplete understanding of the term, often results in drawing inferences that standardized definition of data governance does not exist. These elements have been discussed in detail in the third book in this series: Data Governance Success.

Christopher Butler, Chief Data Officer, HSBC UK, states that data proliferation, use, and ease of sharing data, data privacy, increasing data regulation, and businesses' understanding of how data can be used to drive growth is expanding exponentially. It is the successful merging of these areas within an existing organizational structure or effectively changing the organization's structure to take advantage of data that provides a challenge. He believes that the lack of success of many organizations to do this successfully brings about the opinion that no standard definition around data governance exists. However, this is not the case (Butler). There are quite a few perceptions and misunderstandings around the term data governance, the most common misconception being that data governance is synonymous with data quality. These have led to organizations defining data governance differently. This is reinforced by Jill Dyché, Principal, Jill Dyché, LLC. In her interview, when questioned about the evolution of data governance, she stated:

> I started working with clients in data governance back in the mid-2000s. Back then, every client defined data governance differently. Some saw it as a central board of decision-makers. Others considered it to be synonymous with data quality. Now there's much more consensus on what data governance is and how it should be deployed.

A number of data management practitioners, information management technology vendors and researchers have defined data governance in different ways that add depth to the general understanding of the term. Table 3.1 outlines the different definitions of data governance found in literature.

Table 3.1 Data governance definitions—literature review

DAMA International defines data governance as "planning, oversight, and control over management of data and the use of data and data-related resources" (DAMA International 2012)
Maria Villar and Theresa Kushner define data governance as "a management program that treats data as an enterprise asset: a collection of corporate policies, standards, processes, people, and technology essential to managing critical data to a set of goals"
John Ladley defines data governance as "the organization and implementation of policies, procedures, roles, and responsibilities, which outline and enforce the rules of engagement, decision rights, and accountabilities for effective data management" (Ladley 2012)
Sunil Soares defines data governance as "the formulation of policy to optimize, secure, and leverage information as an enterprise asset by aligning the objectives of multiple functions"
Steve Sarsfield defines data governance as "a set of processes ensuring important enterprise data assets are formally managed, guaranteeing data can be trusted by putting people in charge of fixing and preventing issues with data so that the organization has fewer negative events as a result of poor data" (Sarsfield 2009)
Robert Seiner defines data governance as "the execution and enforcement of authority over the management of data and data-related assets" (Seiner 2007)
David Loshin defines data governance as "the specification of a set of information policies that reflect business needs and expectations, along with the processes for monitoring conformance to those information policies" (Loshin 2008)
Cohen (2006) defines data governance as "the process by which a company manages the quantity, consistency, usability, security, and availability of data"
D. Newman and D. Logan (2006) define data governance as "the collection of decision rights, processes, standards, policies, and technologies required to manage, maintain and exploit information as an enterprise resource" (Newman and Logan 2006)
Tony Fisher defines data governance as "the process for addressing how data enters the organization, who is accountable for it, and how—using people, processes, and technologies—data achieves a quality standard that allows for complete transparency within an organization" (Fisher 2009)
Thomas (2006), of the Data Governance Institute, states that data governance, "refers to the organizational bodies, rules, decision rights, and accountabilities of people and information systems as they perform information-related processes." She goes on to state, "data governance sets the rules of engagement that management will follow as the organization uses data"
Joshua Jones defines data governance as "a framework that lays out decision rights, accountabilities, and formal management procedures for information-related processes. Its purpose is to ensure that high-quality data that has usability, security, and integrity exists throughout an organization. Governance describes the who, what, when, why, and how of data management and data-related decisions" (Jones 2018)
Aaron Jones from the MDM Institute defines data governance as "the formal orchestration of people, processes, and technology to enable an organization to leverage data as an enterprise asset"
Jill Dyché and Evan Levy define data governance as "the organizing framework for establishing strategy, objectives, and policies for corporate data" (Dyché and Levy 2006)
National Institute of Standards and Technology (NIST) defines data governance as "a set of processes that ensures that data assets are formally managed throughout the enterprise. A data governance model establishes authority and management and decision-making parameters related to the data produced or managed by the enterprise" (NIST)

(continued)

Table 3.1 (continued)

As per Anne Marie Smith, data governance is "the planning, oversight, and control over the management of data and the use of data and data-related resources, and the development and implementation of policies and decision rights over the use of data. It is the foundational component of an enterprise data management or enterprise information management program" (Smith 2013)
Khatri and Brown (2010) define data governance as "refers to who holds the decision rights and is held accountable for an organization's decision making about its data assets"
Zeljko Panian defines data governance as "the processes, policies, standards, organization, and technologies required to manage and ensure the availability, accessibility, quality, consistency, auditability, and security of data in an organization" (Panian 2010)
OMES Data Governance Program Office defines data governance as "an organizational process and a structure. It establishes responsibility for data, organizing program area staff to collaboratively and continuously improve data quality through the systematic creation and enforcement of policies, roles, responsibilities, and procedures"
As per Price Waterhouse Coopers (PWC), data governance is defined as "is all about building the capability of the organization through effective leadership, policy and culture and behaviors to better manage, protect and exploit your data to achieve your business objectives" (PWC Singapore 2015)
The IBM Data Governance Council (IBM Software Group 2007) defines data governance as "the political process of changing organizational behavior to enhance and protect data as a strategic enterprise asset"
Capgemini defines data governance as "Theoretically, data governance encompasses the systematic and formal management of any service or process that is required for effective information management. But realistically, businesses prioritize and sponsor only those initiatives that are mandated by regulations or provide a clear return on investment" (Capgemini 2013)
Morgan Templar defines data governance as "the discipline and science of understanding, controlling, and measuring the quality and flow of information throughout your business" (Templar 2017)
As per Oracle, "data governance is the formulation of policy to optimize, secure, and leverage information as an enterprise asset by aligning the objectives of multiple functions" (Oracle 2015)
In another definition by Oracle, "data governance is the specification of decision rights and an accountability framework to encourage desirable behavior in the valuation, creation, storage, use, archival and deletion of data and information. It includes the processes, roles, standards, and metrics that ensure the effective and efficient use of data and information in enabling an organization to achieve its goals" (Sun et al. 2011)
First Sans Francisco Partners defines data governance as "the organizing framework for establishing strategy, objectives, and policy for effectively managing corporate data. It consists of the processes, policies, organization, and technologies required to manage and ensure the availability, usability, integrity, consistency, audit ability, and security of your data"
Michele Goetz defines data governance "as a strategic business initiative that determines and prioritizes the financial benefit the data brings to organizations as well as mitigates the business risk of poor data practices and quality" (Goetz 2015)
Wikipedia defines data governance as "the people, processes, and information technology required to create consistent and proper handling of an organization's data across the business enterprise. Goals may be defined at all levels of the enterprise, and doing so may aid in acceptance of processes by those who will use them"

(continued)

Table 3.1 (continued)

Informatica defines data governance as "the functional coordination and definition of processes, policies, standards, technologies, and people across the organization to manage data as a corporate asset. This enables managing the availability and controlled growth of accurate, consistent, secure, and timely data for better decision making, reduced risk, and improved business processes" (Informatica 2013)
Experian defines data governance as "a process to ensure data meets precise standards and business rules as it is entered into a system. Data governance enables businesses to exert control over the management of data assets. This process encompasses the people, process, and technology that is required to ensure that data is fit for its intended purpose" (Experian)
Data Governance Australia defines data governance as "the formal orchestration of people, processes, and technology to enable an organization to leverage data as an enterprise asset" (McCarney 2018)
Roe (2017) defines data governance as "a collection of practices and processes which help to ensure the formal management of data assets within an organization"
Margaret Rouse defines data governance as "the overall management of the availability, usability, integrity, and security of the data employed in an enterprise. A sound data governance program includes a governing body or council, a defined set of procedures, and a plan to execute those procedures" (Rouse)
Anthony Chalker, Managing Director Proviti defines data governance as "the specification of decision rights and an accountability framework to encourage desirable behavior in the valuation, creation, storage, use, archiving and deletion of information. It includes the processes, roles, standards, and metrics that ensure the effective and efficient use of information in enabling an organization to achieve its goals" (Chalker 2014)
Dr. Elizabeth Pierce, Chair of the Information Science Department, the University of Arkansas at Little Rock, defines data governance as "specifying the decision rights and accountability framework to encourage desirable behaviors in the use of Data. Data governance design lays out the decision-making structures, alignment processes, and communication approaches that enable the strategic objectives for data and its quality to be implemented and to monitor how well these strategic objectives are being achieved" (Pierce 2007)

There are several aspects of data governance. These aspects include organizational bodies or groups, roles, accountabilities, rules, policies, processes, decision rights, metrics, controls, and other enforcement methods,. When individuals in organizations allude to data governance, they may be referring to one or more of these aspects.

The likening of data governance to any one of these aspects reminds me of the famous parable of six blind men and the elephant, as illustrated in Fig. 3.5. In short, six blind men were asked to touch an elephant and describe it. Each touched a different part and described it differently. The blind man who touched the elephant's trunk said it was a snake; the one who touched the tail said it was a rope; the one who felt the leg said it was a tree trunk; the one who touched the ear said it was like a fan; the one who touched the body said that it was a wall; and the one who touched the tusk said that it was a spear.

Data governance programs can vary significantly, depending on their focus (on compliance, data integration, or master data management, etc.). Regardless of the "flavor" of governance, however, every program will have the following components:

Fig. 3.5 Six men and the blind elephant

- organizational structure, decision rights, accountabilities, and responsibilities related to data,
- rules of engagement,
- policymaking and enforcement through guidelines, processes, standards, rules, metrics, and controls,
- data issue resolution,
- to monitor/enforce compliance while providing ongoing support to data stakeholders (DataGovernance.com).

Data governance is *NOT*

- a project that has a finite end date, but an ongoing long-term effort that needs to be systematic and sustained to be successful,
- a technology or tool,
- a feature,
- a change management program (Sun et al. 2011),
- data management,
- data quality,
- master data management,
- data ownership,
- data stewardship,
- data modeling,

- data warehousing or extract, transform and load data (ETL) (Sun et al. 2011),
- database design (Sun et al. 2011), and
- database management and administration (Sun et al. 2011).

While each of the above is affected by or connected to the data governance program, data governance addresses more than the above elements, and each of the above elements has features beyond data governance. Data governance is a cross functional function that ensures that data is treated as an enterprise asset. It manages and resolves risks and issues associated with data, and contributes towards a data-driven culture.

Since data governance is cross-functional, it involves stakeholders from different business units, each of whom needs to have clearly defined roles and responsibilities concerning the data. Also active collaboration is needed to have a consensus regarding data decisions and a common understanding related to different aspects of data (for example, data security, data quality, and so on). For these reasons, the implementation of data governance needs an operating model and organization structure. While the data governance operating model and organization structure vary from organization to organization, common groups/bodies involved in data governance at different levels are Executive Steering Committee, Data Governance Council, Data Stewardship Council, Data Governance Office (DGO), and Information Technology Partners.

Data has a life cycle starting with the creation/acquisition, processing/transformation for different uses, storage, and maintenance in different systems and transfer across different systems, archival, and purging. There are different functional titles such as data producers, data publishers, data consumers, data stewards, data owners, and data custodians that data governance uses in association with data and who have a connection to the different aspects of the data life cycle. For example, data producers are associated with the creation of data. Data consumers use data for a specific purpose, define business and quality requirements about the data, and verify that the data is fit for use.

Figure 3.6 gives a brief description of the organization bodies and functional titles that constitute the people aspect of data governance. Data governance also needs policies, guidelines, processes, standards, rules, controls, enforcement mechanisms, and metrics to oversee the management of data and prevent and resolve data issues and mitigate risks associated with the data. While data policies are a high-level statement that defines what should be done to data in line with the organizations' vision for data, guidelines and processes outline how to implement the data policies and the steps and methods to govern the data.

Rules and standards are documented agreement related to data for implementation of data policies and as to how the data is represented and formatted, naming conventions, metadata requirements, data security standards, and business rules, and data quality rules. Metrics and controls are monitoring mechanisms to ensure adherence to standards and rules.

Organizations have a lot of data, a large number of data producers and consumers, and a large number of policies, processes, guidelines, standards, rules,

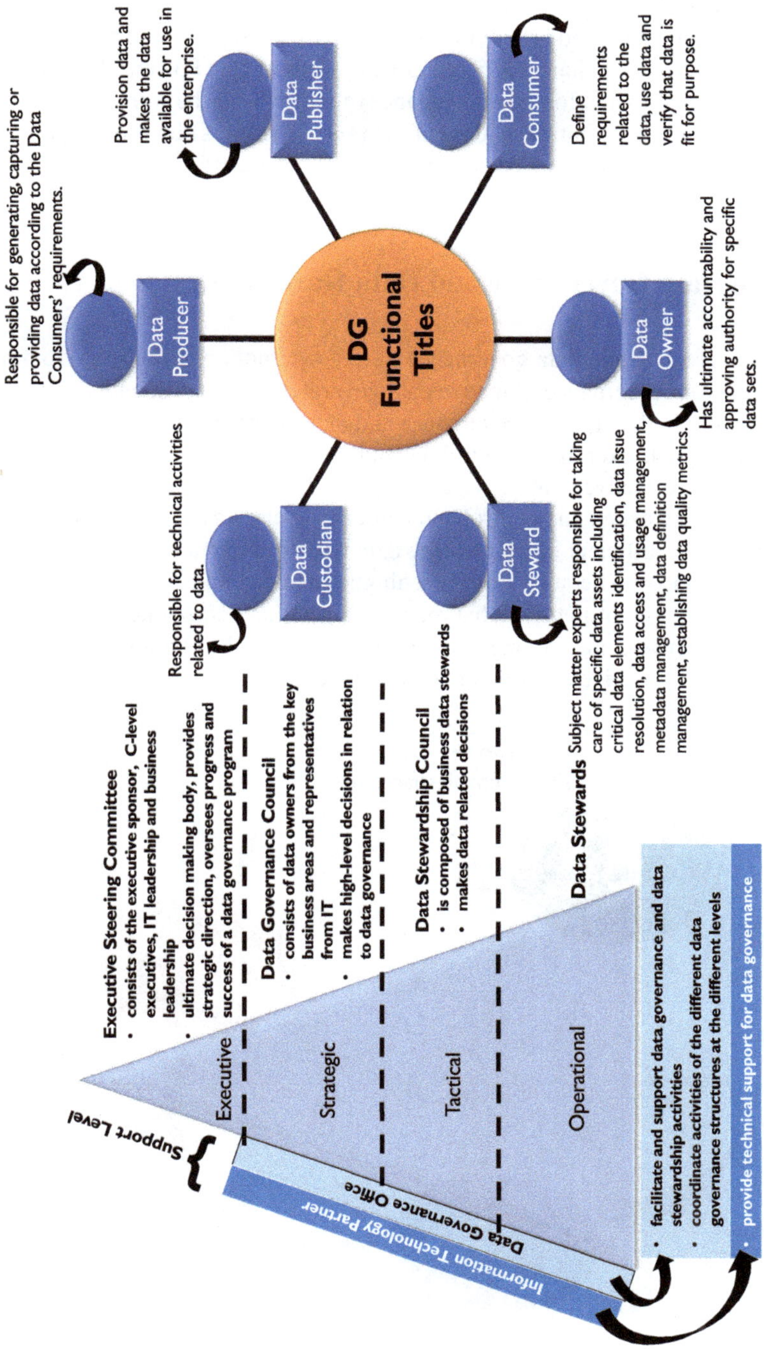

Fig. 3.6 People aspect of data governance

and controls. Managing all these aspects manually requires considerable time and effort. Tools and technology act as a data governance enabler by enabling automation of some of these elements and improving efficiency. However, it should be noted that data governance is not a technical program. In the words of Dan Power, "Technology won't ever be a silver bullet for data governance (about 80% of the work will be with organizational, cultural, political and business process matters) but don't ignore it altogether; it can be a powerful "force multiplier," (Power 2015).

3.5 Corporate Governance and Data Governance

Corporate governance and data governance have a symbiotic relationship. To have effective corporate governance, effective data governance is crucial, and vice versa.

At the heart of both corporate governance and data governance is control. Corporate governance is the exercising control and authority over corporate management. However, to have effective control, you need good data to support your case. Data governance is about exercising control and authority over data assets. Organizations which do not have sufficient or trustworthy data will not be in a position to control data effectively and maybe at risk of breaching regulations and being penalized. It is a common principle of many global regulations that any corporate management must have an effective control framework. Figure 3.7 shows the relationship between good quality data, effective control, and compliance.

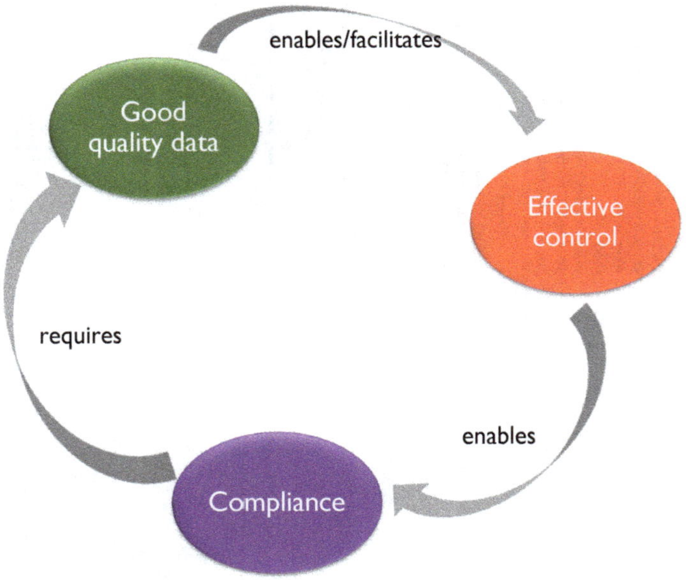

Fig. 3.7 Good quality data, effective control, and compliance

For example, the 2002 Sarbanes-Oxley Act, states:
SEC. 404. MANAGEMENT ASSESSMENT OF INTERNAL CONTROLS.

(a) RULES REQUIRED—The Commission shall prescribe rules requiring each annual report to contain an internal control report, which shall—

   (1) state the responsibility of management for establishing and maintaining an adequate internal control structure and procedures for financial reporting;

The corporate governance code, created by the Financial reporting council in 2014, states in their basic principles of corporate governance (Principle B, effectiveness) that:

> The board should be supplied in a timely manner with information in a form and of a quality appropriate to enable it to discharge its duties.

Solvency II is another example of this kind of regulation in relation to data and governance and is covered under Article 41 (Parkinson 2016). This states:

> That system shall at least include an adequate transparent organizational structure with a clear allocation and appropriate segregation of responsibilities and an effective system for ensuring the transmission of information.

Data governance ensures that good quality data is available for corporate governance. However, for data governance to effectively support corporate governance, an organization's data strategy needs to align with the corporate strategy and the data vision, mission, and goals in the data strategy need to be aligned to achieve the corporate vision and mission, and corporate goals. These elements enable data goals to be framed from a corporate governance perspective. Also, the data requirements of the board in terms of key performance indicators might help in strategic decision making.

Hence, corporate governance feeds into data governance, the corporate vision, mission, goals, corporate strategy, and the board's data requirements. In contrast, data governance feeds good quality data and processes to sustain data quality into corporate governance.

3.6 IT Governance Versus Data Governance

Both IT governance and data governance are subset disciplines of corporate governance. IT governance is often confused with data governance and vice versa because both IT governance and data governance have related and interdependent goals. Both disciplines aim to improve the company's assets to generate better business value. Table 3.2 compares IT governance and data governance.

The IT processes in COBIT IT governance to achieve financial reporting is based on internal control of the COSO framework (Hawkins et al. 2003), introduced in

Table 3.2 IT governance versus data governance

IT governance	Data governance
Manages technology assets (http://www.acolyst.com/data-it-governance/)	Manages data assets (http://www.acolyst.com/data-it-governance/)
IT governance is application and IT infrastructure-centric	Data governance is data-centric
IT governance consists of following established frameworks and best practices to gain the most leverage and benefit out of IT investments and support the accomplishment of business objectives (Smallwood 2014a)	Data governance consists of the processes, methods, tools, and techniques to ensure that data is of high quality, reliable, and unique (not duplicated), so that downstream uses in reports and databases are more trusted and accurate (Smallwood 2014b)
IT governance ensures proactive and effective management of IT investments	Data governance ensures proactive and effective data management
IT governance ensures that the IT infrastructure aligns with business objectives and utilized cost-effectively (Luftman 2004)	Data governance ensures that data management efforts align with business objectives
IT governance enables the organization to manage risks for the IT Infrastructure	Data governance enables the organization to manage risks that are specific to the data itself

1992 by the Committee of Sponsoring Organizations of the Treadway Commission, to provide a management framework for internal controls. The five COSO framework internal control components are: Control Environment, Risk Assessment, Control Activities, Information and Communication, and Monitoring. There is a relationship between data and each of these five components as follows (adapted from Cheong and Chang 2007; Marinos 2004b):

- Concerning the 'Control Environment' component, data quality acts as a foundational platform for internal control, where data management is a process, and the outcome is data quality, and hence it is important.
- With the 'Risk Assessment' element, the COSO framework requires risks to be identified for the business to achieve pre-specified objectives. The data quality affects the extent to which risks can be identified for achieving the pre-specified objectives.
- "Control activities" are procedures and policies which are aligned with business objectives and for executing risk mitigation strategies. With the 'Control Activities' element, data symbolizes the means of control, and to be managed, data need to be measured. Control activities involve the production of reports from enterprise data to measure the achievement of pre-specified objectives.
- About the "Information and Communications' component, data needs to be of high quality. That is, data should be accessible, timely, current, consistent, valid, accurate, free of duplicates, and complete for effective communication and decision making.

- The "Monitoring" process depends on high-quality data, which is current, accurate, timely, and secure. This is imperative for measuring actual performance against suitable operation ranges of a particular activity.

Hence, the success of the COBIT framework depends on the quality of the underlying enterprise data, which is also supported by Marinos (2004b), who states that "data quality is the hidden assumption behind COSO." Hence, to ensure effective IT governance, an organization should have an effective data governance framework in place to oversee the effective management of data assets.

3.7 Governance, Risk, and Compliance (GRC)

The acronym ***"GRC"*** is an umbrella term for *"Governance, Risk Management, and Compliance" or "Governance, Risk, and Compliance"* which came into existence in the early 2000s (Wheeler et al. 2016). Due to numerous high profile corporate scandals and financial failures, regulations for improved compliance were introduced with the U.S. Sarbanes Oxley Act of 2002 being the initial driver. This, in turn, necessitated organizations to improve their internal control and governance processes.

GRC is a strategy for organizational governance in a manner that risks are effectively managed while staying compliant with regulations. As the name suggests, GRC comprises of three constituents:

- G for Governance,
- R for Risk, and
- C for Compliance.

Figure 3.8 shows an overview of GRC with the arrow showing the exchange of information amongst the three components: Governance, Risk, and Control (Salido and Voon 2010).

- ***Governance***: Ensuring that organizational activities are aligned in a manner that supports the organization's business and compliance goals, establishes policies and processes, organization's risk appetite, decision making authorities and accountabilities for activities, associated risks and outcomes, and a framework for evaluating performance.
- ***Risk***: Relates to the identification of the potential risk areas and identification, analysis, addressing the associated risks and monitoring the risks by establishing risk thresholds in a way that supports the organization's business and compliance objectives goals. Controls play an important role in managing risks. Risk requirements change with changes like the implementation of new business processes and models and deployment of new technologies. Hence risk impacts of these changes need to assessed and addressed. Existing controls might need to be modified or new controls might need to be implemented to address the risk impacts. Controls are usually costly and may adversely impact business performance; hence the

Fig. 3.8 GRC overview (adapted from Salido et al. 2010)

right level of effective controls should be introduced to optimize performance within compliance and risk threshold boundaries.

- ***Compliance***: Making sure that organizational activities are operated in a way that actions and behaviors adhere to laws, internal obligations, and external regulations, as well as appropriate processes and tools, are in place to assess and validate compliance. Meeting compliance involves creating appropriate controls, as well as auditing those controls to ensure they are working as intended. Compliance mitigates risks. Compliance requirements change with regulation changes, and hence, processes and activities to ensure compliance need to be revised, and new controls need to be put in place, if necessary, to meet the changing regulations. For example, changes in the global regulatory environment and industry standards.

GRC developed as an integrated approach, because of the interdependencies amongst these components. GRC ensures that an organization operates under self-imposed rules, tolerable risk levels, and external regulations. Risk management and compliance without governance would result in informal rules, informal roles and responsibilities, and informal policies, which will likely be chaotic. Benefits of a good GRC strategy are improved and informed decision-making concerning risk mitigation, effective controls, cost reduction, elimination of silos, and greater collaboration among business units and departments.

3.8 Data Governance and GRC

Data needs to be managed effectively to be compliant with internal and external regulations. The risks related to data like unauthorized access, data hacking, confidential data exposure, data breaches, and frauds resulting from low-quality data need to be managed effectively. Also, good quality data is a key factor in the success of GRC. Data needs to be accurate, current, timely, consistent, trustworthy, auditable, and secure, to satisfy risk and compliance processes and requirements. This is where data governance comes in. Data governance establishes roles and responsibilities, accountabilities, standards, rules, processes, and practices so that data is managed effectively and efficiently to ensure high-quality data, which in turn feeds into the risk and compliance processes.

3.9 Tying Data Governance with the Different Disciplines of Governance

Corporate vision and mission and governance sit at the top. Corporate vision and mission drive the corporate strategy with corporate governance being a part of the corporate strategy. **Corporate strategy** is hierarchically the highest **strategic plan** of the organization. It defines the corporate strategic objectives as to what the organization is set to achieve and be, and the directions, commitments, decisions, and activities to achieve those goals. The data vision and mission drive the data/data quality strategy with data governance being a part of the organization's data/data quality strategy. An organization's data vision, mission, data/data quality strategy should align with its corporate vision, mission and corporate strategy, respectively. There is an exchange of information between corporate governance and the various subset disciplines of corporate governance—operational governance, financial governance, risk governance, HR governance, security governance, IT governance, and data governance. That is, the corporate governance feeds into the various subset disciplines and vice versa. However, for corporate governance as well as any of the subset disciplines of governance to function effectively, they need facts to base their decisions and strategy on, and data governance facilitates the provision of good quality data (facts that are not distorted) as well as good quality metadata (what the data means or data about data) to the different sub-disciplines, namely operational governance, financial governance, risk governance, HR governance, security governance, and IT governance that is needed by their respective business units to govern their respective assets effectively (see Fig. 3.9).

It is important to note, though, that data governance itself does not provide the data but helps provision good quality data and metadata by setting up adequate policies, processes, standards, and accountabilities, and responsibilities around the management of data.

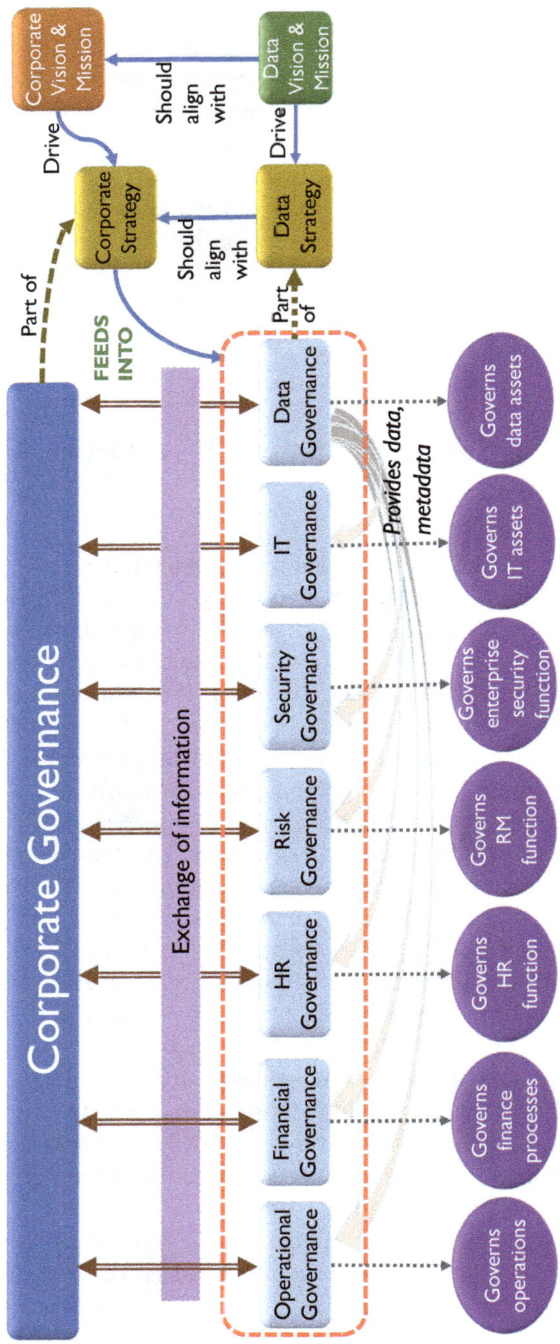

Fig. 3.9 Tying the different disciplines of governance (adapted from Parkinson February 2016)

3.10 Concluding Thoughts

This chapter discusses the various subset disciplines under corporate governance, namely, operational governance, financial governance, HR governance, security governance, risk governance, IT governance, and data governance.

For any of the subset disciplines of governance to function effectively, they need facts. These facts form the foundation for their decisions and strategy. Data represents facts, and data governance facilitates the provision of good quality data and good quality metadata (that is, data about data) to operational governance, financial governance, risk governance, HR governance, security governance, and IT governance by setting up the processes, accountabilities, and responsibilities throughout the lifecycle of the data.

Both information technology and data management have evolved significantly in the past few decades. The evolution of hardware and software technologies have resulted in the capture, processing, and storage of electronic data of various sizes and variety at different speeds. As data has evolved, new threats and issues have also emerged and expanded. Management and governance practices have also evolved to deal with these changes in the world of data. This is reinforced by Dr. John R. Talburt, Acxiom Chair of Information Quality at the University of Arkansas at Little Rock, and Lead Consultant for Data Governance and Data Integration with Noetic Partners Inc., when he illustrates the same with an example in his interview statement: "Data governance is clearly growing in adoption and maturing in practice. Having been around for many years, I have seen many shifts in paradigms around computing. For example, one of the great benefits of the data warehousing movement was not just to create business intelligence (BI), but it also brought focus on the problems of information and data quality."

There is still a lot of work to be done. This was also highlighted by Laura Sebastian-Coleman, Data Quality Lead, Finance Data Governance Center of Excellence at CVS Health in response to a question regarding interesting thoughts/advice. "I think we are heading into a new era of data governance. The existing assumptions and models for data governance have not produced the results that people have expected. There is still a limited understanding of the risks associated with data in organizations. Until we face these more squarely, I do not think we can progress in the ways that we want to. Even the idea that data is an asset still has not resulted in improvements in how we manage data. People are more focused on technology than on data. So…be aware as you engage in this work, that it takes a lot of persistence and a willingness to try new things."

In response to a question regarding interesting thoughts/advice that Andres Perez, Data Architect, IRM Consulting Ltd. Co., would like to share with the audience of the book about data governance, this is what he had to say: "Data governance is not new, data stewardship is not new…..What is new is their widespread practice leading to better understanding, shared knowledge, and better tools (collaboration tools, catalogs, etc.). We are beginning to break away from (using Auguste Comte's evolutionary model) the ***theological stage***, in which data was thought as mythically

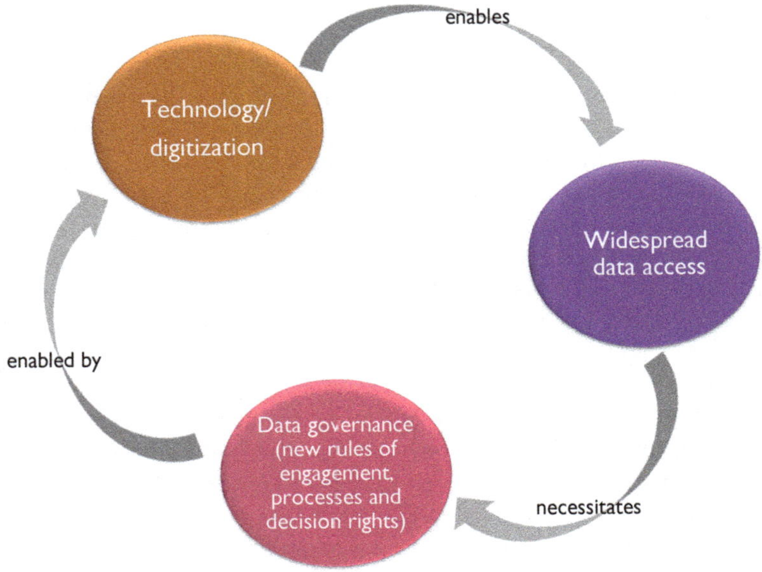

Fig. 3.10 Technology, widespread data access, and data governance

conceived and explained in dogmatic ways. Now we are in the ***metaphysical stage*** in which data is perceived as an obscure force resulting from the "natural ways in which organizations operate" explained in obscure ways. I hope that what we learn next will lead us into the ***positive stage*** in which all abstract and obscure forces are discarded, and data phenomena are explained by real and objective (value based) means."

Also what is new, is widespread access of data, enabled by technology and digitization, which necessitates adequate data governance with new rules of engagement, processes, and decision rights that are enabled by technology as shown in Fig. 3.10.

While data governance is an extremely important data management function, it has several interacting elements that tie together the other data management functions, with several drivers, use cases, and benefits that steer organizations to embrace and implement data governance.

There are also significant challenges and perceptions that can act as blockers or inhibitors for a data governance program and key success factors that can drive and help sustain the data governance program.

Compliance is one of the biggest drivers of data governance and will be discussed in Chap. 5. While a lack of compliance can have adverse impacts on an organization, a singular focus on compliance is not a solution. Organizations need to focus on performance too to ensure sustainability. Both compliance and performance are resource-intensive, and meeting both compliance and performance objectives requires a robust strategy, execution plan as well as discipline by understanding the business goals for performance, the business constraints, regulatory and compliance constraints,as well as how these performance goals, compliance requirements, and constraints tie up together. If compliance and performance requirements can go hand in hand, it is a double win.

However, if ensuring compliance adversely impacts performance, steps should be taken to make sure that rules, laws, and regulations are correctly interpreted, compliance risks are assessed, and right level of controls (as opposed to excess controls) are implemented to achieve compliance while balancing performance (Lustig 2014).

References

Barrett P (2001) Corporate governance—more than good management. In: Proceedings of the CPA South Australia annual congress 2001—"Riding the Next Wave", Adelaide, Australian National Audit Office

Capgemini (2013) Data governance for financial institutions. https://www.capgemini.com/wp-content/uploads/2017/07/data_governance_for_financial_institutions.pdf Accessed 7 July 2018

Chalker A (2014) Data governance overview, Proviti, GEEK Week, ISACA Atlanta. https://www.isaca.org/chapters3/Atlanta/AboutOurChapter/Documents/GW2014/Implementing%20a%20Data%20Governance%20Program%20-%20Chalker%202014.pdf Accessed 7 July 2018

Chen W (2010) Brief history of data governance, data governance blog, Magnitude.com, 25 June 2010. https://magnitude.com/blog/a-brief-history-of-data-governance/ Accessed 7 June 2018

Cheong LK, Chang V (2007) The need for data governance: a case study. In: ACIS 2007 proceedings, p 100. http://aisel.aisnet.org/acis2007/100 Accessed 7 June 2016

Chisholm M (2010) Definitions in data management: a guide to fundamental semantic metadata. Design Media, Print

Cohen R (2006) BI strategy: what's in a name? Data governance roles, responsibilities and results factors. DM Rev. http://www.dmreview.com/article_sub.cfm?articleId=1057220 Accessed 12 Jan 2007

DAMA International (2012) DAMA-DMBOK2 framework. https://dama.org/sites/default/files/download/DAMA-DMBOK2-Framework-V2-20140317-FINAL.pdf Accessed 25 June 2020

De Haes S, Van Grembergen W (2005) IT governance structures, processes and relational mechanisms: achieving IT/Business alignment in a major Belgian financial group. In: Proceedings of the 38th Hawaii international conference on system sciences

Dreibelbis A, Hechler E, Milman I et al (2008) Chapter 1: Introducing master data management. Enterprise master data management: an SOA approach to managing core information. Pearson Education, pp 1–3. ISBN 9780132704274

Dyché J, Evan L (2006) Customer data integration: reaching a single version of the truth. Copyright © 2006, SAS Institute Inc, Cary, North Carolina, USA

Experian, Data Governance. https://www.edq.com/glossary/data-governance/

First Sans Francisco Partners, The first step in master data management. http://www.sourcemediaconferences.com/MDMCanada12/Malinga_Sustainable%20Governance%20Tutorial.pdf Accessed 8 July 2018

Fisher T (2009) The data asset: how smart companies govern their data for business success. Wiley, Hoboken, NJ

Gartner, Gartner IT glossary. https://www.gartner.com/it-glossary/it-governance/ Accessed on 7th July 2018

Goetz M (2015) Data governance and data management are not interchangeable, forrester blog, 1 Sept 2015. https://go.forrester.com/blogs/15-09-11-data_governance_and_data_management_are_not_interchangeable/ Accessed 7th July 2018

Hawkins K, Alhajjaj S, Kelley S (2003) Using COBIT to secure information assets. J Gov Financ Manag 52(2):22–33

Informatica (2013) Metadata management for holistic data governance, Informatica White Paper, July 2013. http://egovstandards.gov.in/sites/default/files/Metadata%20Management%20for%20Holistic%20Data%20Governance%20%28Year%20of%20Publication%202012%29.pdf Accessed 14 Jan 2019

International Finance Corporation (2013) International finance corporation control environment toolkit: risk governance, model risk management committee charter, Sec 2.1.25 (internal document on file with IFC)

International risk governance council. What is risk governance?. https://irgc.org/risk-governance/what-is-risk-governance/ Accessed 10 July 2018

ISACA (2008) IT governance institute, enterprise value: governance of IT investments, the Val IT Framework, Version 2.0

i-Sight (2018) A practical guide to data privacy laws by country. https://i-sight.com/resources/a-practical-guide-to-data-privacy-laws-by-country/ Accessed 29 July 2018

ITGI (2001) Board briefing on IT governance, IT governance Institute. www.ITgovernance.org and www.isaca.org Accessed 7 Aug 2003

ITGI (2003) Board briefing on IT governance. http://www.itgi.org

IT governance Ltd. https://www.itgovernance.co.uk/it_governance

IT governance Ltd[a], ISO 38500—International standard for corporate governance of IT. https://www.itgovernance.eu/en-ie/iso-38500-ie

Jones J (2018) Data governance and master data management (MDM/DG) glossary, Strategy Wise, 31 Jan 2018. https://strategywise.com/mdmdg/ Accessed 7 July 2018

Khatri V, Brown CV (2010) Designing data governance. Commun ACM 53(1):148–152. https://doi.org/10.1145/1629175.1629210

Ladley J (2012) Data governance: how to design, deploy and sustain an effective data governance program, 1st Edn. The Morgan Kaufmann Series on Business Intelligence, Paperback. ISBN: 9780124158290

Lake P, Crowther P (2013) Concise guide to databases, undergraduate topics in computer science. © Springer, London. https://doi.org/10.1007/978-1-4471-5601-7_2

Lipman FD, Lipman LK (2006) Corporate governance best practices: strategies for Public, Private, and Not-for-Profit Organizations. Print ISBN: 9780470043790 | Online ISBN:9781119197195 | https://doi.org/10.1002/9781119197195

Loshin D (2008) Master data management. Morgan Kaufmann

Luftman JN (2004) Managing the information technology resources. Pearson Prentice Hall, New Jersey

Lustig M (2014) Compliance at speed. O'Reilly Media, Inc. ISBN: 9781491909874

Mahanti R (2019). Data quality: dimensions, measurement, strategy, management and governance. ASQ Quality Press, Milwaukee

Marinos (2004b) Data quality: the hidden assumption behind COSO, DM review, Oct, p 12

McCarney P (2018) Data governance—what is it? Data governance Australia. http://datagovernanceaus.com.au/data-governance-what-is-it/

Mercer Human Resource Consulting (Mercer), Inc (2003) Why HR governance matters. http://www.adirelab.es/site/wp-content/uploads/2016/04/COMPLIANCE-WhyHRGovernanceMatters.pdf Accessed 7th July 2018

Newman D, Logan D (2006) Governance is an essential building block for enterprise information system. Gartner Research

National Institute of Standards and Technology (NIST), Glossary-Data Governance. https://csrc.nist.gov/Glossary/?term=3846 Accessed 7 July 2018

OED Online Etymology Dictionary. data (n.), https://www.etymonline.com/word/data

OECD (2014) Risk Management and Corporate Governance, p 7

OMES Data governance program office, data governance overview. https://www.ok.gov/cio/documents/DataGovernanceOverview.pdf Accessed 7 July 2018

Oracle (2015) Data governance with Oracle. Defining and implementing a pragmatic data governance process with Oracle metadata management and Oracle data quality solutions, Oracle White Paper, Sept 2015. http://www.oracle.com/us/products/middleware/data-integration/oracle-data-governance-2672357.pdf Accessed 7 July 2018

Otto B (2011) Organizing data governance: findings from the telecommunications industry and consequences for large service providers. Commun Assoc Inf Syst 29, Article 3

Panian Z (2010) Some practical experiences in data governance, World Academy of Science, Engineering and Technology. http://citeseerx.ist.psu.edu/viewdoc/download?doi=10.1.1.190.6948&rep=rep1&type=pdf Accessed 8 July 2018

Parkinson J (2016) Corporate governance and data governance. Are you really In Control? 9 Feb 2016 https://www.linkedin.com/pulse/corporate-governance-data-you-really-control-john-parkinson/ Accessed 1 Dec 2019

Pearce G (2017) Align data governance with board governance imperatives, The Data Administration Newsletter. https://tdan.com/align-data-governance-with-board-governance-imperatives/21355 Accessed 1 Dec 2019

Phillips B (2012) IT governance for CEOs and members of the board, p 18

Pierce EM (2007) Designing a data governance framework to enable and influence IQ strategy. In: Proceedings of the MIT 2007 information quality industry symposium. http://mitiq.mit.edu/IQIS/Documents/CDOIQS_200777/Papers/01_08_1C.pdf Accessed 12 Dec 2018

Power D (2015) Secrets of successful data governance, information governance. Blog-Eckerson.com, Jan 2015. https://www.eckerson.com/articles/secrets-of-successful-data-governance Accessed 10 Dec 2015

PWC Singapore (2015) Building a data enabled business. https://isca.org.sg/media/2238702/introduction-to-data-governance-isca-oct-2017.pdf Accessed 7 July 2018

Renn O (2008) Risk governance: coping with uncertainty in a complex world. Earthscan, London

RMIA (2015) Body of knowledge domain 6 governancehttps://irgc.org/risk-governance/what-is-risk-governance/

Roe C (2017) What is Data Governance, Data Versity, Dec 2017. http://www.dataversity.net/what-is-data-governance/ Accessed 7 June 2018

Rouse M Data governance definition. www.whatis.techtarget.com Accessed 9 Apr 2017

SABPP (2017) SA Board for people practices, May 2017, fact sheet. http://sabpp.co.za/wp-content/uploads/2017/05/Fact-Sheet_May_v003-with-active-links.pdf Accessed 7 July 2018

Salido J,Voon P (2010), A guide to data governance for privacy, confidentiality, and compliance, Jan 2010, Microsoft White Paper

Sarsfield S (2009) The data governance imperative. IT Governance Publishing. ISBN-10: 1849280126

SAS (2017) The SAS® Data governance framework: a blueprint for success. https://www.sas.com/content/dam/SAS/en_us/doc/whitepaper1/sas-data-governance-framework-107325.pdf Accessed 14 Dec 2017

Sebastian-Coleman L (2013) Measuring data quality for ongoing improvement: a data quality assessment framework. Morgan Kaufmann. Print. The Morgan Kaufmann Series on Business Intelligence

Seiner RS (2007) How to build & implement "Non-Invasive Invasive©*" data governance & data stewardship programs. CDI-MDM Spring Summit MDM Spring Summit, San Francisco San Francisco USA, 25 Mar 2007. http://www.sourcemediaconferences.com/CDISP07/pdf/Seiner_Bob.pdf Accessed 20 July 2019

Smallwood R (2014a) Information governance: Concepts, strategies and best practices. Wiley, NJ

Smallwood R (2014b) Defining the differences between information governance, IT governance, & data governance, AIIM community blog. http://community.aiim.org/blogs/robert-smallwood/2014/08/18/defining-the-differences-between-information-governance-it-governance–data-governance Accessed 10 June 2018

Smith AM (2013) Foundations of data governance, EW Solutions, 20th July 2013. https://www.ewsolutions.com/foundations-data-governance/ Accessed 7 July 2018

Soares S, Franco JM (2017) Data governance & privacy ocmpliance: 16 practical steps towards GDPR compliance with talend, talend white paper, July 8, 2017 Copyright © 2017 Information Asset, LLC, and Talend, Inc. https://info.talend.com/rs/talend/images/WP_EN_TLD_Talend_Outlining_PracticalSteps_GDPR_Compliance.pdf Accessed 2 May 2019

Sun, Helen; Oracle, 2011, Enterprise Information Management: Best Practices in Data Governance, An Oracle Whitepaper on Enterprise Architecture. http://egovstandards.gov.in/sites/default/files/oea-best-practices-data-gov-400760.pdf Accessed on 7 July 2018

TageTik, Financial governance. https://www.tagetik.com/au/glossary/financial-governance Accessed 1 Jan 2018

Templar M (2017) Get governed: building world class data governance programs. Ivory Lady Publishing

The Institute of Operational Risk Operational risk governance. https://www.ior-institute.org/sound-practice-guidance/operational-risk-governance Accessed 30 Apr 2018

Thomas G (2006) Alpha males and data disaster. Brass Cannon Press, USA

Vagadia B (2014) Enterprise governance, management for professionals. Springer, edition 127, number 978-3-642-38589-6

Vitria (2008) Formalizing operational governance: ensuring the well-managed enterprise. https://www.vitria.com/pdf/WP-Formalizing-Operational-Governance.pdf Accessed 20 May 2020

Villar K, Kushner T, Data governance fundamentals. www.elearningcurve.com.

Wheeler JA (2016) GRC redefined: gartner adopts an integrated risk management view, 17 Oct 2016. https://blogs.gartner.com/john-wheeler/grc-redefined-gartner-adopts-an-integrated-risk-management-view/ Accessed 7 June 2018

Wolter R, Haselden K (2006). The what, why, and how of master data management, Nov 2006. Microsoft Corporation. Archived from the original on 14 July 2017. Accessed 13 Dec 2017

Wiki1. https://en.wikipedia.org/wiki/Corporate_governance_of_information_technology

Wiki-Data. https://en.wikipedia.org/wiki/Data

Wiki-Master. https://en.wikipedia.org/wiki/Master_data

Chapter 4
Regulations

> Good regulation should be conducive to business and to customer protection. – Jamie Dimon

Abstract Regulations are rules or laws created by government or authorities in order to control behaviors or way activities are conducted. They are essential for proper functioning of the society and economy. Over the period of time regulations have evolved with changes in technology. In this chapter we discuss some of the standards and regulations.

4.1 Introduction

Regulations are rules or laws created by government or authorities in order to control behaviors or way activities are conducted. They are essential for proper functioning of the society and economy. Over the period of time regulations have evolved with changes in technology. In this chapter we discuss some of the standards and regulations.

The following sections provide a brief overview of the below standards and regulations:

- Health Insurance Portability and Accountability Act (HIPAA)
- General Data Protection Regulation (GDPR)
- Know Your Customer (KYC), Anti-Money Laundering and Counter-Terrorism Financing Act 2006 (AML/CTF Act)
- Foreign Account Tax Compliance Act (FATCA)
- Basel Committee on Banking Supervision's standard number 239 (BCBS 239)
- Common Reporting (COREP) and Financial Reporting (FINREP)
- Payment Card Industry Data Security Standard (PCI DSS).

R. Mahanti, *Data Governance and Compliance*,
https://doi.org/10.1007/978-981-33-6877-4_4

4.2 Health Insurance Portability and Accountability Act (HIPAA)

HIPAA compliance encompasses fulfilling the requirements of the Health Insurance Portability and Accountability Act of 1996, its subsequent amendments, and any associated regulation such as the Health Information Technology for Economic and Clinical Health (HITECH) Act.

The Health Insurance Portability and Accountability Act (HIPAA) is a United States legislation that was enacted by the United States Congress and signed by President Bill Clinton on August 21, 1996.

HIPAA sets the standard for protecting sensitive patient data and sets the national standards for the protection, confidentiality, security, and transmissibility of personal health information (PHI).

This act imposed a new set of requirements on following U.S. organizations involved with the provision of health care.

- Health Plans—individual and group plans that provide or pay the cost of health care (for example, health maintenance organizations (HMOs), health insurance companies, group health plans, Medicare, and Medicaid).
- Health Care Providers—providers of health care or medical services or supplies (e.g., institutional providers such as hospitals and nursing homes; non-institutional providers such as physicians, dentists, and psychologists) who electronically transmits health information in connection with transactions for which HHS has adopted a standard.
- Health Care Clearinghouses—Public or private entities that process another entity's health care transactions from a standard format to a non-standard format, or vice versa, such as billing services, repricing companies, community health management information systems, and value-added networks.

Health Plans, Health Care Providers and Health Care Clearinghouses are known as covered entities. A person or entity that performs certain functions or activities that involve the use or disclosure of protected health information on behalf of, or provides services to, a covered entity, the person or entity is considered as a "business associate" (TOC_HIT). A business associate also needs to comply with the HIPAA act. The failure to comply with HIPAA can result in civil and criminal penalties.

HIPAA does not provide protections for all health information. HIPAA applies only to PHI. De-Identified Health Information is health information that does not identify an individual, and for which there is no reasonable basis to believe that it can be used to identify an individual. Once PHI is de-identified in accordance with the HIPAA Privacy Rule, it is no longer PHI, or subject to the HIPAA Rules, and, thus, may be used and disclosed by the covered entity or business associate for any purpose without restriction by HIPAA (but remains subject to any other applicable laws) (TOC_HIT).

HIPAA consists of the following rules.

4.2.1 HIPAA Privacy Rule

HIPAA Privacy Rule addresses the use and disclosure of individuals' health information in all forms, for example, print, oral, and electronic format. It permits the disclosure of personal health information needed for patient care and other important purposes. Individually identifiable health information also known as PHI (Protected Health Information) is information, including demographic data that relates to such personal information such as name, address, birth date, social security number, address, and past medical history. HIPAA requirements mandate that this type of information must be protected.

PHI includes information that relates to all of the following:

- The individual's past, present, and future health (physical or mental) or condition,
- The provision of health care to the individual, and
- The past, present, or future payment for the provision of health care to the individual.

4.2.2 HIPAA Security Rule

HIPAA Security Rule specifies a series of administrative, physical, and technical safeguards for covered entities to use to assure the confidentiality, integrity, and availability of electronic protected health information. The HIPAA Security Rule protects a subset of information covered by the HIPAA Privacy Rule. This subset of information is referred to as e-PHI. The Security Rule applies to electronic protected health information (e-PHI) that a covered entity creates, receives, maintains or transmits in an electronic format. It defines and regulates the standards, methods, and procedures related to the protection of electronic PHI on storage, accessibility, and transmission. The Security Rule does not apply to PHI transmitted orally or in writing.

4.2.3 HIPAA Breach Notification Rule

The HIPAA Breach Notification Rule requires U.S. organizations involved with the provision of health care to notify affected individuals, U.S. Department of Health & Human Services (HHS), and, in some cases, the media of a breach of unsecured PHI. Generally, a breach is an impermissible use or disclosure under the Privacy Rule that compromises the security or privacy of PHI.

4.3 General Data Protection Regulation (GDPR)

The General Data Protection Regulation (GDPR) is comprised of 173 Recitals and 99 Articles. Article 33 (1) of the General Data Protection Regulation (GDPR) requires data controllers to notify the competent supervisory authority in case of a personal data breach, without undue delay, and within 72 hours after having become aware of the personal data breach, unless the breach is unlikely to put the rights and freedoms of natural persons at risk. Article 33 (2) of the GDPR states that if a processor becomes aware of a personal data breach processed on behalf of the controller, it must notify that controller without undue delay. Hence, GDPR applies to 'controllers' and 'processors.' A controller determines the purposes and means of processing personal data. A processor is responsible for processing personal data on behalf of a controller.

As compared to its precursor regulation, the Data Protection Directive (95/46/EC), the GDPR empowers data protection authorities with more inspective and enforcement authorities as well as the authority to impose more substantial fines. There are two tier of fines involved, which are outlined in article 83 of the GDPR act (as shown in Table 4.1) with the upper limit of financial penalties at €20 million or 4% of global annual turnover, whichever is greater. One of the overriding principles outlined in the GDPR act is that the penalties should be "effective, proportionate, and dissuasive."

The GDPR sets out the following key principles summarized in Fig. 4.1 (ICO):

- **Lawfulness, fairness and transparency**

Personal data must be processed lawfully, fairly, and in a transparent manner in relation to the data subject.

The lawful bases for processing are set out in Article 6 of the GDPR. At least one of these must apply whenever you process personal data:

Table 4.1 Article 83 of GDPR—Two tiers of fines

Tier of fines	Conditions
Up to €10 million or 2% of total worldwide annual turnover, whichever is greater	**For violations including**: failure to report breaches to the data processing authority (DPA) or to the individual concerned (data subject); and failure to appoint a data protection officer (DPO)
Up to €20 million or 4% of total worldwide annual turnover, whichever is greater	**For violations including**: consent breaches; non-compliance with data access, correction, erasure, and portability rules; and the illegal transfers of personal data to non-compliant EU countries or organizations

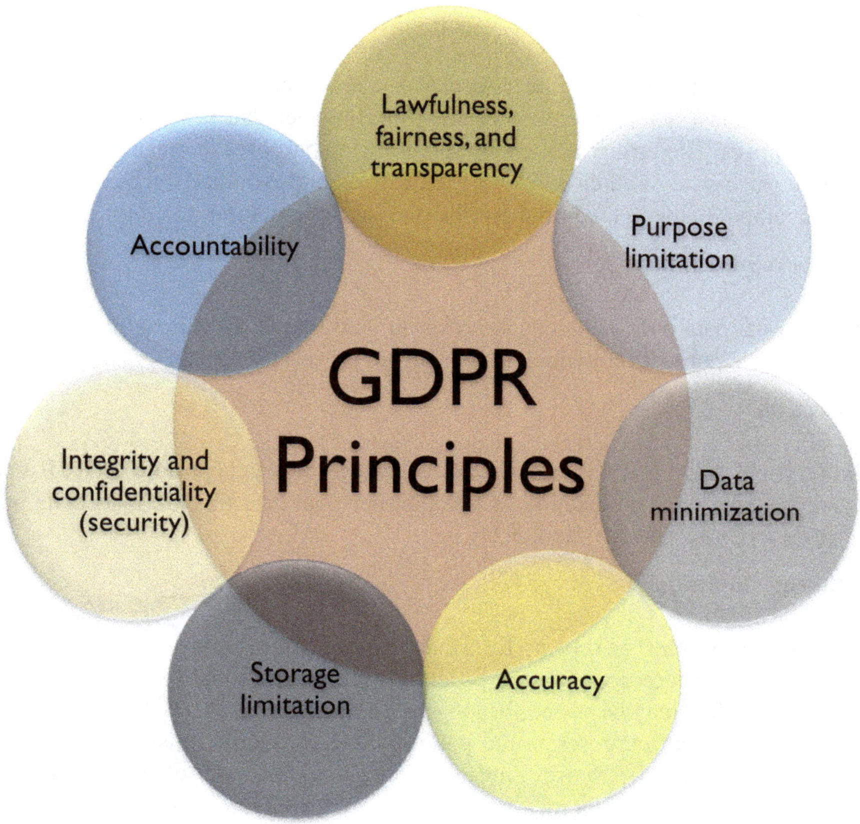

Fig. 4.1 GDPR principles

(a) **Consent**: the individual has given clear consent for you to process their personal data for a specific purpose.
(b) **Contract**: the processing is necessary for a contract you have with the individual, or because they have asked you to take specific steps before entering into a contract.
(c) **Legal obligation**: the processing is necessary for you to comply with the law (not including contractual obligations).
(d) **Vital interests**: the processing is necessary to protect someone's life.
(e) **Public task**: the processing is necessary for you to perform a task in the public interest or for your official functions, and the task or function has a clear basis in law.
(f) **Legitimate interests**: the processing is necessary for your legitimate interests or the legitimate interests of a third party unless there is a good reason to protect the individual's personal data which overrides those legitimate interests. (This cannot apply if you are a public authority processing data to perform your official tasks.)

- **Purpose limitation**

The personal data may only be collected for specified, explicit, and legitimate purposes. It must not be further processed in a manner that is incompatible with those purposes. The further processing for archiving purposes in the public interest, scientific, historical research purposes or statistical purposes shall not be considered to be incompatible with the initial purpose ('purpose limitation').

- **Data minimization**

The personal data must be adequate, relevant, and limited to what is necessary in relation to the purposes for which those data are processed.

- **Accuracy**

The personal data must be accurate and, where necessary, kept up to date. Every reasonable step must be taken to ensure that personal data that are inaccurate are either erased or rectified without delay.

- **Storage limitation**

The personal data must be kept in a form that permits identification of data subjects for no longer than is necessary for the purposes for which the personal data are processed. The personal data may be stored for longer periods in so far as it will be processed solely for archiving purposes in the public interest, scientific, historical research purposes or statistical purposes subject to the implementation of the appropriate technical and organizational measures required by the GDPR in order to safeguard the rights and freedoms of individuals ('storage limitation').

- **Integrity and confidentiality (security)**

The personal data must be processed in a manner that ensures appropriate security of that data, including protection against unauthorized or unlawful processing, against accidental loss, destruction or damage, using appropriate technical or organizational measures.

- **Accountability**

The controller is responsible for and must be able to demonstrate compliance with the Data Protection Principles.

Article 25 of GDPR defines the concept of "data protection by design and by default," also known as "privacy by design and by default."

Privacy by design: The enterprise must protect privacy throughout the operations lifecycle, from initial process, and systems design through service end-of-life and data deletion.

Privacy by default: The enterprise must ensure that, by default, only the required amount of personal data is collected and processed. The user should not have to opt out from giving extra information. The enterprise cannot gather more information "just in case" it might want to use it later.

The GDPR provides the following rights for individuals (ICO) as shown in Fig. 4.2:

1. **The right to be informed**

The right to be informed covers some of the key transparency requirements of the GDPR. It is about providing individuals with clear and concise information about what organizations do with their personal data. Articles 13 and 14 of the GDPR specify what individuals have the right to be informed about and this is called 'privacy information.'

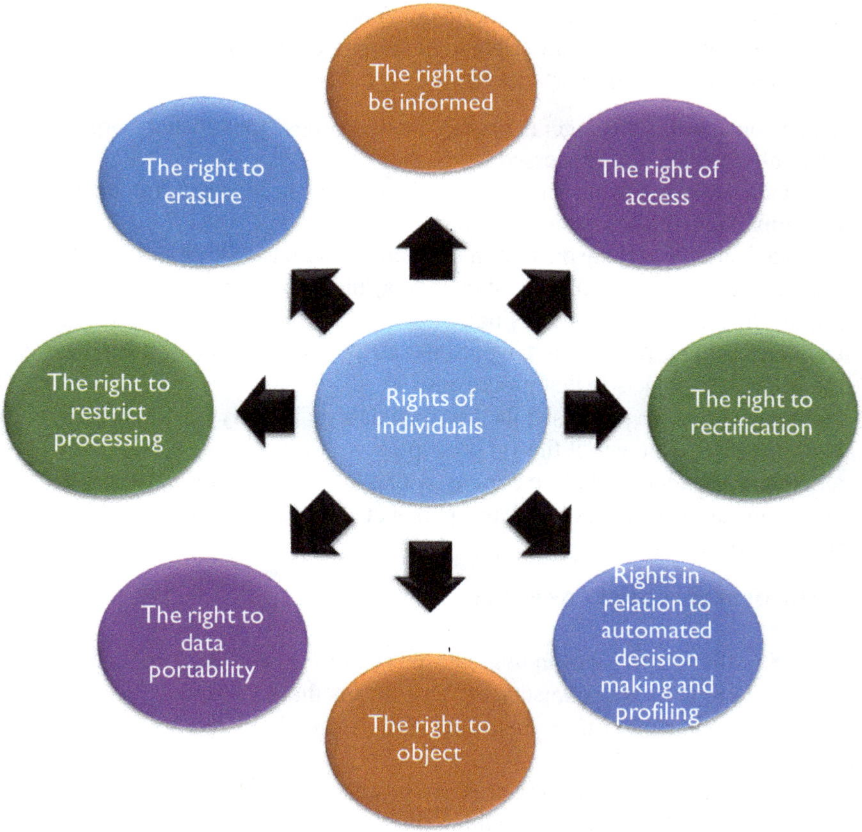

Fig. 4.2 Rights of individuals under GDPR

Using an effective approach can help you to comply with other aspects of the GDPR, foster trust with individuals, and obtain more useful information from them. Getting this wrong can leave you open to fines and may also lead to reputational damage.

2. **The right of access**

The right of access, commonly referred to as subject access, gives individuals the right to obtain a copy of their personal data as well as other supplementary information.

3. **The right to rectification**

Under Article 16 of the GDPR, individuals have the right to have inaccurate personal data rectified.

4. **The right to erasure**

Under Article 17 of the GDPR, individuals have the right to have personal data erased. This is also known as the 'right to be forgotten.' The right is not absolute and only applies in certain circumstances as follows:

- the personal data is no longer necessary for the purpose for which the organization originally collected or processed it;
- an organization is relying on consent as its lawful basis for holding the data, and the individual withdraws their consent;
- an organization is relying on legitimate interests as a basis for processing, the individual objects to the processing of their data, and there is no overriding legitimate interest to continue this processing;
- an organization is processing the personal data for direct marketing purposes, and the individual objects to that processing;
- an organization has processed the personal data unlawfully (i.e., in breach of the lawfulness requirement of the 1st principle);
- an organization has to do it to comply with a legal obligation; and
- an organization has processed the personal data to offer information society services to a child.

5. **The right to restrict processing**

Article 18 of the GDPR gives individuals the right to restrict the processing of their personal data in certain circumstances. This means that an individual can limit the way that an organization uses their data. This is an alternative to requesting the erasure of their data.

Individuals have the right to restrict the processing of their personal data where they have a particular reason for wanting the restriction. This may be because they have issues with the content of the information an organization holds or how an organization has processed their data. In most cases, an organization will not be required to restrict an individual's personal data for an indefinite period but will need to have the restriction in place for a certain period of time.

6. **The right to data portability**

The right to data portability gives individuals the right to receive personal data that they have provided to a controller in a structured, commonly used, and machine-readable format. It also gives them the right to request that a controller transmits this data directly to another controller.

7. **The right to object**

Article 21 of the GDPR gives individuals the right to object to the processing of their personal data. This effectively allows individuals to ask you to stop processing their personal data. The right to object only applies in certain circumstances. Whether it applies or not, depends on the purposes for processing and the lawful basis for processing.

8. **Rights in relation to automated decision making and profiling**

Automated individual decision-making is a decision made by automated means without any human involvement. Automated individual decision-making and profiling can lead to quicker and more consistent decisions. But if they are used irresponsibly, it may result in significant risks for individuals. The GDPR provisions are designed to address these risks.

The GDPR says that profiling is:

"Any form of automated processing of personal data consisting of the use of personal data to evaluate certain personal aspects relating to a natural person, in particular, to analyze or predict aspects concerning that natural person's performance at work, economic situation, health, personal preferences, interests, reliability, behavior, location or movements."

Automated individual decision-making does not have to involve profiling, although it often will do.

The GDPR restricts an organization from making solely automated decisions, including those based on profiling, that have a legal or similarly significant effect on individuals.

4.4 Know Your Customer (KYC), Anti-Money Laundering and Counter-Terrorism Financing Act 2006 (AML/CTF Act)

KYC is an acronym for "Know Your Customer," a term used for customer identification process, and involves the determination of the true identity and beneficial ownership of accounts, source of funds, the nature of customer's business, equanimity of operations in the account in relation to the customer's business which in turn helps the financial institutions such as banks to manage their risks wisely.

The objective of the KYC guidelines is to prevent the data lying with banks, from being used, intentionally or unintentionally by criminal elements for money laundering. Money laundering is the process of conversion of illegally derived funds into assets that appear to be legitimate to disguise their illegal origins and returning the same to the launderer.

The Anti-Money Laundering and Counter-Terrorism Financing Act 2006 (AML/CTF Act) regulates bullion, gambling, financial, and remittance (money transfer) services, where the conversion, transfer of physical, and electronic forms of money are vulnerable to money laundering and financing of terrorism.

The Anti-Money Laundering and Counter-Terrorism Financing Amendment Act 2017 (AML/CTF Amendment Act) was passed by both Houses of Parliament on 7 December 2017, and received Royal Assent on 13 December 2017. The AML/CTF Amendment Act introduces three key changes:

- Greater regulatory powers and functions for the AUSTRAC CEO.
- Digital currency exchange providers are now subjected to the AML/CTF regime.
- Some regulatory relief for the industry.

These regulations require collecting and verifying certain 'know your customer' (KYC) information about a customer's identity. They apply equally to individuals and non-individuals. These regulations extend in some circumstances to beneficial owners and controlling persons and has many elements, including client identification, identity verification, transaction monitoring, screening for sanctions, blacklists, reporting suspicious activity, and ongoing management of customers' information.

Entities which are required to report transactions to AUSTRAC are called "reporting entities." The "Reporting entities" are required to report transactions to AUSTRAC. The transactions which must be reported include (Deakin and Liu 2018):

- cash transactions of A$10,000 or more, or foreign currency of that value,
- international funds transfer instructions, either into or out of Australia, of any amount, and,
- suspicious transactions of any kind, being transactions the dealer may reasonably suspect of being part of tax evasion or crime, or might assist in a prosecution.

Australia's cash controls require travelers to report to AUSTRAC when they carry $10,000 or more (or equivalent in a foreign currency) of cash (or equivalent) into or out of Australia, which can be done on forms available from the Border Force at airports and sea ports. The Border Force attempts to detect evasion of this requirement. Airlines are not liable for what their passengers carry. The cross-border movement of bearer negotiable instruments of any amount must also be reported if requested by a Border Force or a police officer.

4.5 Foreign Account Tax Compliance Act (FATCA)

The Foreign Account Tax Compliance Act (FATCA) is a United States (US) legislation signed into law in March 2010. It came into effect in January 2013, and is aimed at improving compliance with US tax laws, by requiring foreign financial institutions (FFIs) and non-financial foreign entities (NFFEs) to identify and report US citizens, US tax-residents, US corporations, US Residents Account Holders with income and assets overseas to the US Internal Revenue Service (IRS).

The failure to comply with FATCA's requirements will expose such Financial Institutions to a 30% US withholding tax on payments to them from any US sources. An FFI is a foreign financial institution, which is any non-U.S. entity that (Deloitte 2013):

- Accepts deposits in the ordinary course of a banking or similar business;
- As a substantial portion of its business, holds financial assets for the account of others;
- Is engaged (or holding itself out as being engaged) primarily in the business of investing, reinvesting, or trading in securities, partnership interests, commodities, or any interest in such securities, partnership interests, or commodities;
- Is a specified insurance company; or
- Is a holding company or treasury center.

The U.S. final regulations lists following seven indicia of U.S. status and Notice 2011-34 provides details of the required documentation associated with each indicia of U.S. Status (Deloitte 2013):

- A U.S. citizenship or lawful permanent resident (green card) status;
- A U.S. birthplace;
- A U.S. residence address or a U.S. correspondence address (including a U.S. P.O. box);
- A U.S. telephone number (regardless of whether such number is the only telephone number associated with the account holder),
- Standing instructions to pay any amounts from the account to an account maintained in the U.S.;
- An "in care of" address or a "hold mail" address that is the sole address with respect to the client; or
- A power of attorney or signatory authority granted to a person with a U.S. address.

Having one of these indicia does not mean that the account is owned by a U.S. person, only that it must be given closer scrutiny.

The FATCA data requirements go beyond the data collected through existing Know Your Customer (KYC) processes. The Act requires the foreign financial institutions (FFIs) and non-financial foreign entities (NFFEs) to classify customers on a global basis under the U.S. tax principles relying on a range of information such as US Tax Status, US Citizen/Residence status, and place of birth (Nixon 2011).

4.6 Basel Committee on Banking Supervision's Standard Number 239 (BCBS 239)

"BCBS 239" (Basel Committee on Banking Supervision's standard number 239) is a common shorthand term for referring to the paper "Principles for effective risk data aggregation and risk reporting" published in January, 2013, by the Basel Committee on Banking Supervision (BCBS), a global forum and standard setter for regular cooperation and regulation of banks on banking supervisory matters (BCBS 2013).

BCBS 239 was created with an intent to reform the global financial system to avoid a repeat of the global financial crisis of 2008, when many banks were found to have no clear idea of their overall exposures and other key risk metrics because of poor quality data, immature data processes, and reporting practices indicating gaps within the financial system.

Risk data aggregation is the process by which banks collect and process data across all risk types (Credit, Market, Operational, Fiduciary, Interest-Rate, Liquidity, and so on), a quality filter crucial to measure risk. The overall objective of the standard is to strengthen banks' risk data aggregation capabilities and internal risk reporting practices, in turn, enhancing the risk management and decision-making processes at banks.

The standard consists of 14 principles grouped into four related categories aimed at strengthening risk data aggregation and reporting as follows and shown in Fig. 4.3:

Fig. 4.3 BCBS 239 principles

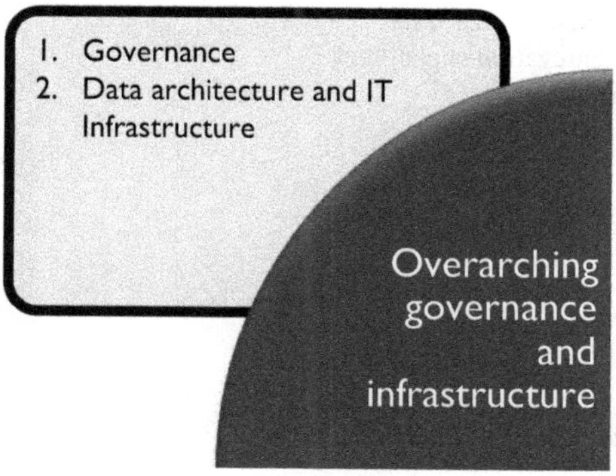

Fig. 4.4 Overarching governance and infrastructure principles

I. Overarching governance and infrastructure;
II. Risk data aggregation capabilities;
III. Risk reporting practices; and
IV. Supervisory review, tools, and cooperation.

A brief description of the 14 basic principles under the respective categories is given below (BCBS 2013).

I. **Overarching governance and infrastructure** (Fig. 4.4)

A bank should have in place a strong governance framework, risk data architecture, and IT infrastructure to ensure compliance with the other principles.

Principle 1

Governance—A bank's risk data aggregation capabilities and risk reporting practices should be subject to strong governance arrangements consistent with other principles and guidance established by the Basel Committee.

Principle 2

Data architecture and IT infrastructure—A bank should design, build, and maintain data architecture and IT infrastructure which fully supports its risk data aggregation capabilities, and risk reporting practices not only in normal times but also during the times of stress or crisis, while still meeting the other principles.

II. **Risk data aggregation capabilities** (Fig. 4.5)

Banks should develop and maintain strong risk data aggregation capabilities to ensure that risk management reports reflect the risks in a reliable way.

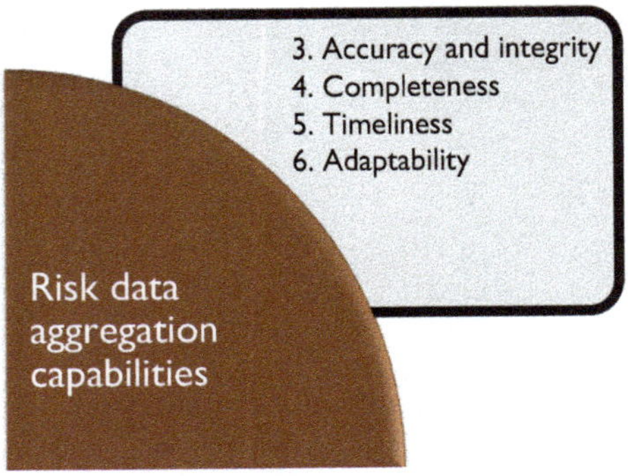

Fig. 4.5 Risk data aggregation capabilities principles

Principle 3

Accuracy and Integrity—A bank should be able to generate accurate and reliable risk data to meet normal and stress/crisis reporting accuracy requirements. The data should be aggregated on a largely automated basis so as to minimize the likelihood of errors.

Principle 4

Completeness—A bank should be able to capture and aggregate all material risk data across the banking group. The data should be available by business line, legal entity, asset type, industry, region, and other groupings, as relevant for the risk in question, that permit identifying, reporting risk exposures, concentrations, and emerging risks.

Principle 5

Timeliness—A bank should be able to generate aggregate and up-to-date risk data in a timely manner while also meeting the principles relating to accuracy and integrity, completeness, and adaptability. The exact timing will be subject to the nature and possible unpredictability of the risk being measured as well as its cruciality to the overall risk profile of the bank. The exact timing will also hinge on the bank-specific frequency requirements for risk management reporting, under both normal and stress/crisis conditions, set based on the features and overall risk profile of the bank.

Principle 6

Adaptability—A bank should be able to produce aggregate risk data to meet a extensive range of on-demand, ad hoc risk management reporting requests, including requests during stress/crisis situations, requests due to fluctuating internal needs, and requests to meet supervisory queries.

III. **Risk reporting practices** (Fig. 4.6)

Accurate, complete, and timely data is a basis for effective risk management. In order to manage risk efficiently, the right information needs to be accessible to the

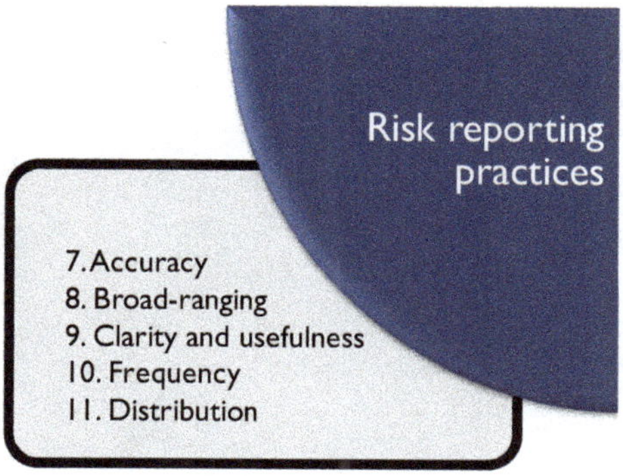

Fig. 4.6 Risk reporting practices principles

right people at the right time. The risk reports based on risk data should be accurate, clear, and complete. They should contain the correct content and be presented to the appropriate decision-makers in a timeframe that allows for an appropriate response. To effectively achieve their objectives, risk reports should comply with the following principles.

Principle 7

Accuracy—Risk management reports should correctly and precisely convey aggregated risk data and reflect risk in an precise manner. Reports should be reconciled and validated.

Principle 8

Comprehensiveness—Risk management reports should cover all substantial risk areas within the organization. The depth and scope of these reports should be consistent with the magnitude and complexity of the bank's operations and risk profile, as well as the requirements of the recipients.

Principle 9

Clarity and usefulness—Risk management reports should convey information in a clear and succinct manner. The reports should be easy to understand yet inclusive enough to enable informed decision-making. Reports should include a suitable balance between risk data, analysis, interpretation, and qualitative explanations. The reports should include meaningful information customized to the recipients' needs.

Principle 10

Frequency—The board and senior management (or other recipients as apt) should set the frequency of risk management report creation and circulation. The frequency requirements should reflect the needs of the recipients, the nature of the risk reported, the swiftness with which the risk can change, and the importance of reports in contributing to sound risk management practices, and effective and efficient decision-making across the bank. The frequency of reports should be augmented during periods of stress/crisis.

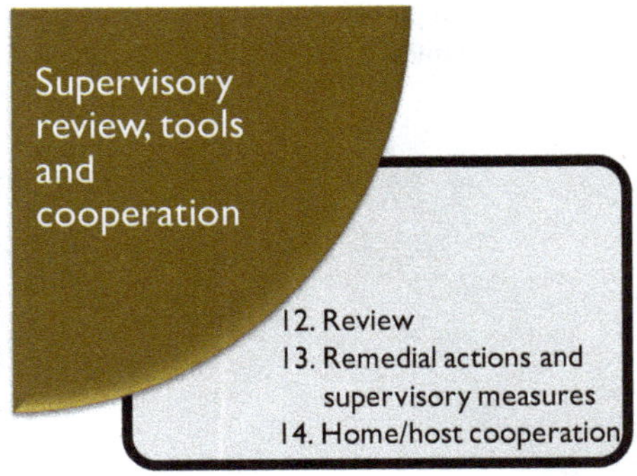

Fig. 4.7 Supervisory review, tools and cooperation principles

Principle 11

Distribution—Risk management reports should be circulated to the pertinent parties while safeguarding that confidentiality is preserved.

IV. **Supervisory review, tools, and cooperation** (Fig. 4.7)

Supervisors will have an imperative role to play in monitoring and providing incentives for a bank's implementation of, and ongoing compliance with the principles. They should also review compliance with the principles across banks to determine whether the principles themselves are attaining their desired outcome, and whether further enhancements are required.

Principle 12

Review—Supervisors should regularly review and assess a bank's compliance with the eleven Principles above.

Principle 13

Remedial actions and Supervisory measures—Supervisors should have (and use) the appropriate tools and resources to implement effective and timely remedial action by a bank to address deficiencies in its risk data aggregation capabilities and risk reporting practices. The supervisors should have the ability to use a range of tools, including Pillar 2.

Principle 14

Home/host cooperation—Supervisors should liaise with the pertinent supervisors in other jurisdictions concerning the supervision and assessment of the Principles, and the implementation of any corrective action if necessary.

One of the biggest challenges that financial organizations face in meeting BCBS 239 is the need and ability to report on consolidated risks on short notice, within hours. To achieve this, the organizations need access to risk data and activity quickly from the large volumes of data that an organization stores.

A structured approach, the right data architecture, the data mart dedicated to storing the risk data, the right technology, governance structures, policies, standards, processes, and controls to ensure that the data captured and stored are of high quality. In addition, a collaborative effort from cross-functional teams comprising of Information Technology, Risk, Compliance, and Finance can help furnish the required data to regulators and minimize risk.

4.7 Common Reporting (COREP) and Financial Reporting (FINREP)

The global financial crisis (GFC) in 2007–2008, resulted in recognition of more detailed financial reporting requirements. A set of international banking regulations were developed by the Basel Committee to promote stability in the international financial system commonly known as Basel III.

Under the Commission's proposals, Capital Requirements Directive (CRD) was divided into two legislative instruments: the Capital Requirements Regulation (the CRR) and the CRD IV Directive.

Through its Capital Requirements Directive (CRD) IV, the European Commission has made it mandatory for the European Banking Authority (EBA) to develop Implementing Technical Standards (ITS) related to supervisory reporting requirements.

Besides the main goal of aligning the prudential requirements across European countries, CRD IV has also made proposals on corporate governance and remuneration. In response to this mandate, the European Banking Authority (EBA) developed the following two frameworks that are applicable across Europe:

- Common Reporting (COREP), and
- Financial Reporting (FINREP).

The objectives of both COREP and FINREP are to enhance the structure of regulatory reporting and empower banks with greater capability to aggregate their risk data; adopt high-quality internal risk reporting practices; and aid the senior management in banks leading to an improved financial and risk decision making, facilitating trend predictions and global assessment of risk by creating a common basis for furnishing regulatory information. Additional objectives are to make regulatory reporting faster and more standardized by establishing a central repository for European banking data, bring the European reporting requirements onto a single common platform, and eliminate the deviations caused by different supervisors in the European Union. Both the frameworks require massive data at a much more granular level and also require mandatory XBRL reporting. COREP contains 28,486 data points across 31 reports, and FINREP contains 6,682 data points across 32 reports (Source: EBA).

4.7.1 Common Reporting (COREP)

COREP is the framework for the Capital Requirements Directive reporting of risk (Basel III) under these five components: Capital Adequacy, Group Solvency, Credit Risk, Operational Risk, and Market Risk.

This came as a solution to the problems faced by credit institutions and investments firms in taking risk-related decisions in a timely manner during the great European financial crisis at an earlier period. COREP was adopted by the Financial Services Authority (FSA) for the UK banking industry from 31 December, 2012. The FSA has mandated that in the UK, COREP will apply to all credit institutions and investments firms who are regulated under IFPRU (the FSA's prudential sourcebook for investment firms).

4.7.2 Financial Reporting (FINREP)

FINREP is a framework given by EBA for reporting financial (accounting) information to the regulator, which will be applicable to all Credit Institutions in the European Union. The banks must now plan to adjust their reporting models and requirements to adhere to the technical standards published by the EBA on 26 July 2013.

While COREP is a capital reporting regime, FINREP is its financial counterpart.

4.8 Payment Card Industry Data Security Standard (PCI DSS)

PCI DSS is the global data security standard administrated by PCI Security Standards Council and adopted by the payment card brands for all entities that process, store, or transmit cardholder data.

PCI DSS was created in 2006 to increase controls around cardholder data to prevent credit card fraud and data breaches. The standard was developed and backed by the major five payment brands (Master Card, Visa, American Express, Discover, and JCB) in response to increasing credit fraud. It covers technical and operational system components included in or connected to cardholder data.

The goal is to establish, maintain, and guarantee that cardholder information remains secure. All merchants and service providers who store, process, and transmit credit card information must comply with the standard. PCI DSS provides 12 high-level requirements organized into six logical groups (as shown in Table 4.2) for securing cardholder data (PCI SSC 2010).

The "Cardholder" or payment data covers information such as the full primary account number (PAN), the cardholder's name, the credit card service code, and the

Table 4.2 PCI DSS high level requirements (*Source* PCI SSC 2010)

Logical groups	PCI DSS requirements
Build and maintain a secure network	1. Install and maintain a firewall configuration to protect cardholder data 2. Do not use vendor-supplied defaults for system passwords and other security parameters
Protect cardholder data	3. Protect stored cardholder data 4. Encrypt transmission of cardholder data across open, public networks
Maintain a vulnerability management program	5. Use and regularly update anti-virus software or programs 6. Develop and maintain secure systems and applications
Implement strong access control measures	7. Restrict access to cardholder data by a business need to know 8. Assign a unique ID to each person with computer access 9. Restrict physical access to cardholder data
Regularly monitor and test networks	10. Track and monitor all access to network resources and cardholder data 11. Regularly test security systems and processes
Maintain an information security policy	12. Maintain a policy that addresses information security for all personnel

expiration date. The sellers are also responsible for protecting sensitive authentication data in the magnetic-stripe data (e.g., CAV2, CVC2, CVV2, CID, PINs, and PIN blocks).

All companies who are subject to PCI DSS standards must be PCI compliant. There are four levels of PCI Compliance and these are based on volume of card transactions an organization processes per year, as well as other details about the level of risk assessed by payment brands (Wiki-PCI-DSS). At a high level, the levels are following:

- Level 1—Over 6 million transactions annually.
- Level 2—Between 1 and 6 million transactions annually.
- Level 3—Between 20,000 and 1 million transactions annually.
- Level 4—Less than 20,000 transactions annually.

The validation of compliance is performed annually or quarterly, (StoreKit) either by an external Qualified Security Assessor (QSA) or by a firm specific Internal Security Assessor (ISA) that creates a Report on Compliance for organizations handling large volumes of transactions, or by Self-Assessment Questionnaire (SAQ) for companies handling smaller volumes (Wiki-PCI-DSS).

4.9 Concluding Thoughts

In this chapter we discuss some of the standards and regulations. In the next chapter we will discuss the impact the regulations have on data, the cost of non-compliance with some real-life examples to drive the message home, the role of data governance in being compliant, regulatory compliance, and data management.

References

Basel Committee on Banking Supervision (BCBS) (2013) Principles for effective risk data aggregation and risk reporting, © Bank for International Settlements 2013. ISBN 92-9197-913-9 (online) https://www.bis.org/publ/bcbs239.pdf. https://www.risk.net/risk-management/5781976/regulators-bristle-at-slow-progress-on-bcbs-239

Deakin M, Liu S (2018) New anti-money laundering and counter-terrorism financing laws, LegalTalk-Insights, PricewaterhouseCoopers https://www.pwc.com.au/legal/assets/legaltalk/new-anti-money-laundering-and-counter-terrorism-financing-laws-300118.pdf

Deloitte (2013) FATCA frequently asked Questions (FAQs): closing the distance

Information Commissioner's Office (ICO). The guide to the general data protection regulation (GDPR). https://ico.org.uk/media/for-organisations/guide-to-the-general-data-protection-regulation-gdpr-1-0.pdf Accessed on 19 May 2019

Nixon D (2011) FATCA: the data challenge, bank systems and technology. United Business Media

Payment Card Industry Data Security Standard (Wiki-PCI-DSS). https://en.wikipedia.org/wiki/Payment_Card_Industry_Data_Security_Standard

PCI Security Standards Council (PCI SCC) (2010) PCI DSS quick reference duide storeKit, "What you need to know about PCI DSS compliance: UK costs & checklist". Retrieved from https://storekit.com/payments/pci-dss/ on December 18, 2018

TOC_HIT. The office of the national coordinator for health information technology guide to privacy and security of health information, Version 1.0 022112, https://www.healthit.gov/sites/default/files/pdf/privacy/onc_privacy_and_security_chapter4_v1_022112.pdf

Chapter 5
Data Governance and Compliance

> It takes less time to do a thing right, than it does to explain why you did it wrong.
> – Henry Wadsworth Longfellow

Abstract One of the main business drivers of data governance is compliance. Compliance is about following laws and rules to prevent wrong things from happening as well as detecting and mitigating incorrect actions of the past. Regulations have made organizations much more aware of the issues regarding the data they hold. With new regulations coming into the picture and old ones being repeatedly revised, it is imperative to plan for the future and to implement best practices in data governance to stay compliant and achieve better performance.

5.1 Data Governance, Regulations, and Compliance

One of the main business drivers of data governance is compliance. According to a 2017 survey by erwin, Inc. and UBM, 60% of organizations said that compliance was driving their data governance initiatives (erwin 2017). Compliance is about following laws and rules to prevent wrong things from happening as well as detecting and mitigating incorrect actions of the past.

The industries in various sectors and geographies are inundated with a variety of complex laws, rules, and regulations, intended for different purposes, the bulk of which require specific standards of compliance to be met. Compliance in this context refers to one or more of the following (also illustrated in Fig. 5.1):

- regulatory compliance,
- contractual compliance,
- compliance with internal requirements,
- meeting reporting standards, and,
- regulatory reporting.

R. Mahanti, *Data Governance and Compliance*,
https://doi.org/10.1007/978-981-33-6877-4_5

Fig. 5.1 Compliance—setting the context

As data continues to evolve as well as grow exponentially, criminals find new ways to break the law. With the evolution of technology, new risks emerge. As a result, new regulations come into the picture, and existing regulations also need to be amended to become more stringent to minimize corporate failures, fraud, operational failures, financial scandals, and ethical issues. Therefore, the need to stay compliant is extremely critical, as shown in Fig. 5.2.

With new regulations coming into the picture and old ones being repeatedly revised, it is imperative to plan for the future, implement best practices in data governance to stay compliant, and achieve better performance.

Since regulations first appeared in the twentieth century, regulators were more focused on auditing paper records in organizations. Post the advent of the internet, with huge amounts of data being electronically stored, data management systems and the flow of data have come under the scrutiny of regulators and auditors. 85 percent to 95 percent of all regulatory evidence is now electronically stored information (Gow 2006).

The general perception is that data governance is all about compliance. This is because compliance is one of the biggest business drivers of data governance, and a lot of organizations implement data governance to achieve data compliance requirements.

Regulations have made organizations much more aware of the issues regarding the data they hold. However, it should be noted that there are many other factors also driving the increase in data governance such as cybersecurity, data privacy, digitalization, and a rise in customers' expectations [Butler].

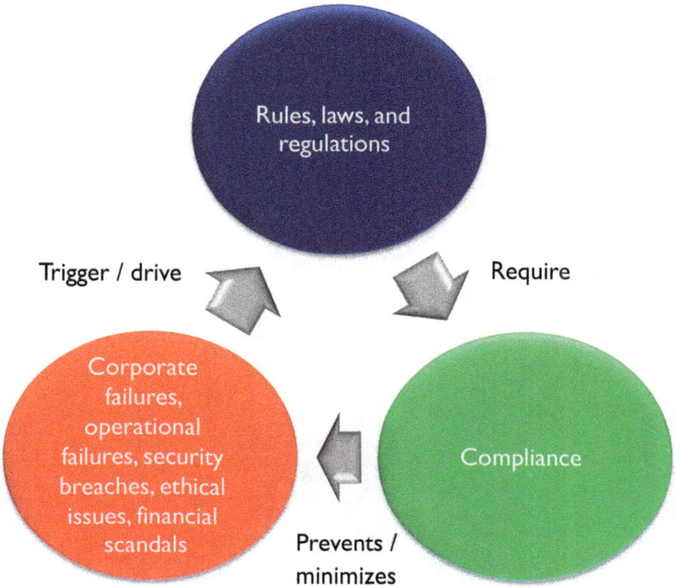

Fig. 5.2 Regulations and compliance

Indeed, you cannot ensure compliance without effective data governance. This is because without effective data governance, you cannot be sure that your organization is fully adhering to laws and regulations. Without complete control over the entire data life cycle, including creating, acquiring, processing, storing, usage, archiving, and purging data, it is not possible to be sure that you are following regulations.

Data governance ensures that processes, policies, rules, standards, metrics, roles, responsibilities, and accountabilities are in place to manage and control data so that regulatory compliances and regulatory reporting requirements are met. Implementing data governance effectively results in an organization's compliance with the demands of regulators by ensuring greater critical data element quality; that is, greater accuracy, completeness, traceability, and data security.

Organizations mistakenly believe that creating data policies in line with regulatory mandates is often sufficient to achieve compliance. An organization often creates a new data policy and adds it to the repository of existing policies, and considers the job done. However, mere creation of data policies will achieve little towards attaining compliance unless it drives the right behaviors and actions. To drive correct and consistent behaviors and actions, the data policies need to be enforced. In addition, the processes, guidelines, standards, controls, rules, data ownership, and data stewardship need to be established in line with the data policies. Figure 5.3 shows the relationship between policies and behaviors, and compliance with rules, laws, and regulations.

To establish, and sustain a strong compliance program, you need to know and understand the specific data requirements around compliance:

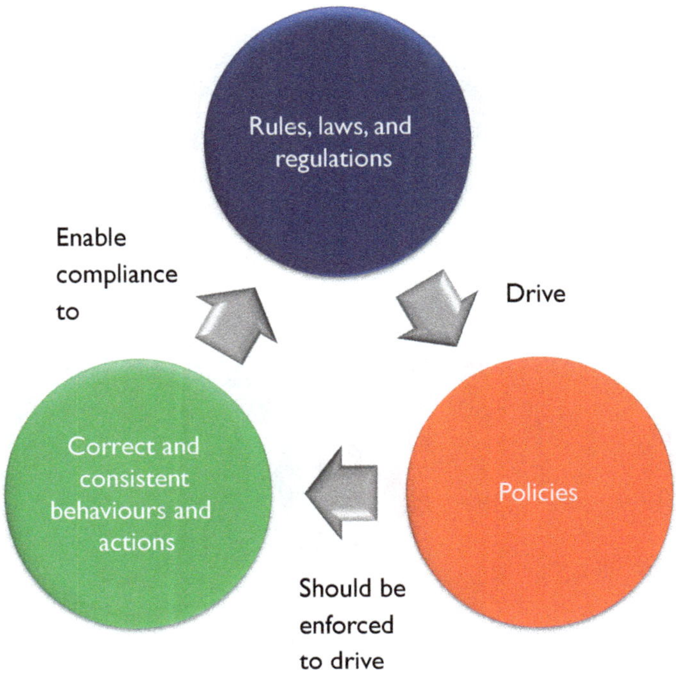

Fig. 5.3 Relationship between policies, behaviors, and compliance to rules, laws and regulations

- Where is the data related to compliance requirements stored? For example, it may be stored in repositories, servers, tables, or files.
- Where is this data sourced from? Can the data be traced back to their original sources?
- How is data integrated or transformed?
- How is data shared and accessed?
- Lastly, how transparent are the data usage practices?

To answer these questions, organizations need an effective implementation plan which allows for data governance to be well-connected with the different data management functions.

Phil Watt, Director, Elait Australia, when questioned about the role of data governance in achieving compliance states that:

> Data governance has a crucial role in compliance. It provides clarity for policies and proof of their application - which is essential for audit. It also provides confidence for the business to act without fear of inadvertently breaking the rules or the law. Similarly, the same approaches, techniques, and technology around data governance for compliance can be applied for profit as they shine a light on the business operations.

The risk of non-compliance can be severe in regulated industries like healthcare, manufacturing, insurance, and finance with hard costs in the form of significant financial penalties, settlement costs, fines, and product recalls (in manufacturing).

Productivity losses and business disruption are other impacts. In addition, there may also be soft costs like difficult-to-repair reputation damage, brand damage, and other undesirable consequences. The risk of non-compliance and associated consequences vary depending on the industry sector, size of the organization, and the risks involved. Some of these regulations include HIPAA, AML, GDPR, BCBS 239, PCI DSS, SoX, OSHA and FSMA GMP.

Regulatory requirements are constantly evolving, with new regulations coming up, and old regulations being revised. For example, FSMA amended Sect. 415 of the Federal Food, Drug, and Cosmetic Act to give the FDA the authority to suspend a food facility's registration if there is a "reasonable probability of [the food it produces] causing serious adverse health consequences or death to humans or animals." The food produced in that facility can no longer be sold if the license is suspended (Nilfisk 2016). Regulatory requirements generally require adequate data protection, data security to ensure data privacy, prevention of fraud or theft, and reporting of certain key data elements at predefined levels of granularity to ensure that the values are accurate, complete, current, consistent, and timely.

You need to be proactive in your efforts around data to meet new as well as changing regulatory requirements. Data Security Management which deals with the adequate protection of data and the role that data governance plays in it will be covered in the second book in this series.

The regulatory drivers can also differ by geographies and countries. This is highlighted by Dr. Stan Rifkin, Director of R&D, Master Systems Inc., in his interview statement:

> In the US, federal sector regulations are probably the governing driver. And in Europe, privacy regulations have shown light on the importance of tending to data.

In this chapter, we discuss the different regulations, the impact they have on data, the cost of non-compliance with some real-life examples to drive the message home, the role of data governance in being compliant, regulatory compliance, and data management.

5.2 Cost of Non-compliance

> If you think cost of compliance is expensive – try non-compliance. – Former US Deputy Attorney General Paul McNulty.
>
> An ounce of prevention is worth a pound of cure. – Benjamin Franklin.

Non-compliance costs are those that result when a company fails to comply with rules, regulations, policies, contracts, and other legal obligations. These include the costs that arise due to business disruptions, productivity losses, revenue losses, damaged brand reputations, fines, penalties, and settlement costs. Other consequences suffered include being forced to participate in remediation programs that include on-site compliance audits/inspections by the appropriate regulatory agency.

Organizations that map to heavily regulated industries, for example, financial services and healthcare, incur the highest compliance costs. These costs are also influenced by the volume of sensitive and confidential data that an organization must secure. While the cost of compliance is high, the cost of non-compliance is significantly higher. Below are some examples and statistics related to the cost of non-compliance:

- HSBC's blatant failure to implement proper anti-money laundering controls facilitated the laundering of at least $881 million in drug proceeds through the U.S. financial system (PricewaterhouseCoopers, 2017).
- The Royal Bank of Scotland was fined $100 m (£61 m, 73 m euros) by US regulators for violating US sanctions against Iran, Sudan, Burma, and Cuba (PricewaterhouseCoopers 2017).
- The Commerzbank AG admits to Sanctions and Bank Secrecy Violations, agrees to forfeit $563 million and pay a $79 million fine (PricewaterhouseCoopers 2017).
- The Financial Crimes Enforcement Network (FinCEN) fined the J.P. Morgan Chase Bank, N.A. $461 million for willfully violating the Bank Secrecy Act (BSA) by failing to report suspicious transactions arising out of Bernard L. Madoff's decades-long, multi-billion-dollar fraudulent investment scheme (PricewaterhouseCoopers 2017).
- BNP Paribas S.A. (BNPP), a global financial institution headquartered in Paris, was sentenced for conspiring to violate the International Emergency Economic Powers Act (IEEPA) and the Trading with the Enemy Act (TWEA) by processing billions of dollars of transactions through the U.S. financial system on behalf of Sudanese, Iranian and Cuban entities subject to the U.S. economic sanctions. BNPP was sentenced to a five-year term of probation, and ordered to forfeit $8,833,600,000 to the United States and to pay a $140,000,000 fine (PricewaterhouseCoopers 2017).
- Barclays bank received a fine of $3.75 million from the US Financial Industry Regulatory Authority (BBC 2013; Martijn et al. 2015).
- The employees of a food and beverage major that was found to have violated the Fair Labor Standards Act (FLSA) in the United States were awarded a $2.5 million settlement (Maindarkar and Thykatt 2014; AOGLLP).
- In January 2013, the UK's Information Commissioners Office fined Sony Computer Entertainment Europe a £250 k fine after it found the Japanese giant guilty of allowing a 'serious breach' of the 1998 Data Protection Act. Sony failed to use up to date security software on its PlayStation Network which allowed hackers to break into its online store, exposing a raft of personal information such as names, addresses, dates of birth, and credit card information to criminals in 2011 (Warman 2013; Glenday 2013).
- In 2018, the Information Commissioner's Office (ICO) fined high-street mobile phone retailer Carphone Warehouse with a £400,000 fine under the 1998 Data Protection Act for failing to adequately protect customer data resulting in a data breach in 2015 (Fearn 2015).

- In October 2016, TalkTalk had been fined a record £400,000 fine by the Information Commissioner's Office (ICO) under the 1998 Data Protection Act for security failings which led to the theft of personal data of almost 157,000 customers (Rodionova 2016).
- Hilton, one of the largest hotel and resort chains in the world, was fined $700,000 for a data breach that caused the information of 350,000 cardholders to be exposed (Burgett 2018).
- The Information Commissioner's Office (ICO) fined Equifax £500,000 for failing to protect millions of UK citizens' personal data during a cyber-attack in 2017. Contact information, email addresses, and credit card information of 15 million UK Equifax customers were compromised in a massive hack on its US parent company, Equifax Inc, between 13 May and 30 July 2017 (Keumars 2018).
- In 2018, the UK Information Commissioner's Office imposed a £500,000 ($908,928) fine against Facebook under the 1998 Data Protection Act, in the wake of the Cambridge Analytica scandal, after allowing third-party developers to access user information without sufficient consent (Reid 2018).
- In 2018, the University of Greenwich was the first university to be fined and was slapped with a £120,000 ($160,000) fine by the UK Information Commissioner's Office, under the 1998 Data Protection Act for a security breach in which the personal data of 19,500 students was placed online (BBC News 2018).
- In 2018, Bupa Insurance Services Limited (Bupa) was fined £175,000 by the Information Commissioner's Office (ICO), under the 1998 Data Protection Act, for failing to have effective security measures in place to protect customers' personal information. Between 6 January, and 11 March 2017, a Bupa employee was able to extract the personal information of 547,000 Bupa Global customers and offer it for sale on the dark web (ICO 2018).
- France's data protection regulator, CNIL, issued a €50 million fine (about $56.8 million USD) fine to Google for failing to comply with GDPR (Price 2019).
- In 2018, a Portuguese hospital was fined €400,000 for misuse of patient records, a German social media service was fined €20,000 for lack of password security, and an Austrian business was fined €4,800 for unlawful CCTV surveillance, under GDPR (Blackmore 2019).
- An HHS administrative law judge upheld an HHS Office for Civil Rights finding, requiring the University of Texas MD Anderson Cancer Center in Houston to pay $4,348,000 in civil penalties for HIPAA violations related to the organization's encryption policies, HHS confirmed June 18. The violations included three data breaches in 2012 and 2013, which exposed health information of more than 33,500 people (Cohen 2018).
- The Singapore Health Services (SingHealth) was fined S$250,000 while Integrated Health Information Systems (IHIS), the IT agency responsible for Singapore's public healthcare sector, was slapped with a S$750,000 fine, for failing to take adequate security measures to safeguard personal data. The oversight had contributed to the July, 2018 cybersecurity attack that compromised personal details of 1.5 million SingHealth patients, and breached their data protection obligations outlined in Singapore's *Personal Data Protection Act* (Yu 2019).

- In May, 2014, the Department of Health and Human Services (HHS) handed out $4.8 million worth of HIPAA fines to the New York and Presbyterian Hospital (NYP) and Columbia University (CU) after they submitted a joint breach report that dated back to September 27, 2010. NYP and CU had violated both the HIPAA Privacy and Security Rules by exposing 6,800 patients' electronic protected health information (ePHI) held on their network, including patient status, vital signs, medications, and laboratory results. NYP paid OCR $3,300,000 and CU had to pay $1,500,000, with both agreeing to complete corrective action plans (CAPs) that include risk analyses, developing risk management plans, revising policies, procedures, staff training, and providing OCR with progress reports (Ouellette 2014).
- The Advocate Health Care Network agreed to pay $5.5 million in 2016 after an investigation showed that it had failed to protect patient data, which led to the loss of 4 million patients' information in 2013 (Spitzer 2018).
- On October 15, 2018, the U.S. Department of Health and Human Services, Office for Civil Rights (OCR) announced a $16 million settlement with Anthem, Inc. over alleged violations of federal privacy and security regulations under the Health Insurance Portability and Accountability Act (HIPAA). The settlement resolved an investigation following a data breach that exposed protected health information of nearly 79 million people (Collins and Lightbourne 2018).
- Since 2005, the U.S. Treasury Department imposed fines of more than $2 billion on a number of foreign banks for violating the U.S. sanctions, including a $617 million criminal fine against ING Bank (ING) in June. Lloyds Bank (LYG), Credit Suisse (CS), Barclays (BCS) were also sanctioned with fines for laundering money for Iran, North Korea, Sudan, and other nations under the U.S. sanctions (Isidore 2012).
- The U.K. bank Standard Chartered which has offices in the United States agreed to pay $327 million in 2012 to U.S. authorities to settle charges for violating international sanctions on transactions with Iran, Burma, Libya, and Sudan. In a statement, Standard Chartered acknowledged that it had handled $24 million of transactions on behalf of entities in Iran and a total of $109 million in the other nations against which the U.S. had imposed sanctions (Isidore 2012).
- As per research conducted by Fenergo, a provider of client lifecycle management solutions for financial institutions, fines amounting to $26 billion were imposed for non-compliance with Anti-Money Laundering (AML), Know Your Customer (KYC), and sanctions regulations in a 10-year period since 2008. The United States accounts for nearly 44% of all global regulatory AML/KYC fines, and 91% of the total value ($23.52 billion). In Europe, 83 fines totaling $1.7 billion were imposed, mostly by the U.K. Financial Conduct Authority (FCA). The findings also reveal that Asia Pacific regulators have levied 79 fines worth almost $609 million, commencing in 2011. The Middle East still lags behind other regions for financial enforcements, recording a total of $9.5 million in the ten years since 2008 (Jaeger 2018).

- On September 4, 2018, Dutch bank ING was levied a fine of $900 million for failing to meet Dutch AML compliance. The investigation, conducted by the Netherlands Public Prosecution Service, discovered that the bank failed to execute policies meant to thwart financial-economic crime. From 2010 to 2016, ING's Dutch branch did not meet due diligence standards when it did not report suspicious transactions in its system (Raval 2018).
- The Reserve Bank of India (RBI) leveled a 30.10 million rupee fine (about $420,000) on Deutsche Bank, for failure to observe Indian KYC and AML regulations in November 2018 (Chandler 2019).
- Financial intelligence regulator Austrac fined gaming giant Tabcorp AUD 45 M (USD 35 M) for non-compliance in March 2017 (Chatterjee 2018).
- According to Boston Consulting Group, banks around the globe have paid approximately USD 321 billion in fines since the financial crisis of 2007 and 2008, as the regulators are more focused than ever on compliance (Chatterjee 2018).
- In 2018, the Financial Industry Regulatory Authority (FINRA) fined Morgan Stanley Smith Barney LLC, one of the chief Wall Street banks, $10 million for violating anti-money laundering rules (AML) for an interval of more than five years (Chandler 2019).
- In August 2018, China's central bank, the People's Bank of China, fined five financial institutions somewhere from $100,000 to $250,000 each for not being compliant with AML laws, including Ping An Bank, Shanghai Pudong Development Bank and the Bank of Communications (Chandler 2019).
- In December, 2018, Latvia's financial regulator imposed a 1.2 million euro charge on Blue Orange Bank for AML noncompliance (Chandler 2019).
- In 2018, Financial Industry Regulatory Authority (FINRA) penalized Swiss bank UBS $5 million for AML non-compliance (Chandler 2019).
- GM incurred $35 million worth of fines for failing to report a defect in its ignition switches that was linked to 13 deaths (Cornerstone on Demand 2014). This 2015 report from Swiss Re shows that 52% of recalls cost food producers more than $10 million each (Nilfisk 2016).
- Following a breach affecting 70 million customers in 2014, retail giant Target suffered a loss of revenue as profits fell by $440 million the following quarter (Merkulov 2017).

As per the "The True Cost of Compliance with Data Protection Regulations" study conducted by Ponemon Institute and Globalscape to determine the full economic impact of compliance activities for a representative sample of 53 multinational organizations (Ponemon Institute and Globalscape 2017; Merkulov 2018):

- Non-compliance costs businesses on an average $14 Million; that is, a 45% increase since 2011.
- For those businesses that do conform, costs averaged around $5.47 million annually—less than fifty percent of the cost of non-compliance.
- The cost of compliance differs considerably depending on the organization's industry sector, ranging from $7.7 million for media to more than $30.9 million for financial services.

- Between 2011 and 2017, healthcare organizations, and technology companies experienced the highest growth in costs at 106% and 99%, respectively. Energy, utilities, and retail companies experienced the lowest growth in total compliance costs at 6 and 40%, respectively.
- Cost of non-compliance is 2.71 times the cost of compliance.
- The deployment of a centralized data governance program reduces total compliance costs by $3.01 million.

5.3 Data and Regulations—The Challenges

Organizations are, by nature, data hoarders and are inundated with increasing volumes of data with each passing day. Organizations are also being subject to an increasing number of complex regulations that continue to evolve and become more stringent in the face of data breaches, fraud, and security incidents. Overall, increasing regulations have had (and will continue to have) a massive impact on data management [Butler].

While both regulations and data are created independently of the other, they tend to have an impact on each other and have added new dimensions of complexity and challenges. Let's look at the complexities that each of these elements (that is data and regulations) bring about, that create challenges for organizations, as summarized in Fig. 5.4.

Organizations need evidence to prove to the regulators that they are compliant and accurate, complete, consistent, and traceable data can serve as evidence. While organizations capture and store exorbitant amounts of data, many organizations have no idea what all data they hold. A very small proportion of the data that an organization possesses is used for business purposes or is subject to regulations. Tony Epler, Chief Data Strategist, PricewaterhouseCoopers, reinforces Leandro DalleMule's and Thomas H. Davenport's statement (DalleMule and Davenport 2017) in his interview and states:

> Cross-industry studies show that on average, less than half of an organization's structured data is actively used in making decisions—and less than 1% of its unstructured data is analyzed or used at all.

A survey conducted by the Compliance, Governance and Oversight Council (CGOC) showed that only 1% of information being retained was subject to legal hold requirements (that is, required to be preserved because it is related to the subject matter of actual or reasonably anticipated litigation or regulatory proceedings) (Baker and Sjoberg 2018).

With large amounts of data stored and replicated in multiple repositories across the organization, locating this 1% of data is like trying to find a needle in a haystack. Also, to be compliant, not only does an organization need to possess the ability to locate the data but also be able to produce the right data at the right level of granularity at the right time, and the data needs to be of high quality (that is, it should be accurate, consistent, and complete).

Fig. 5.4 Data and regulations—challenges

Organizations should also be able to trace the data back to authoritative sources if needed. A data discovery exercise can be made less cumbersome by consulting data domain experts and subject matter experts (SMEs) to identify the locations of high concentrations of the specific data entities or data elements. Data discovery tools can also be used to target these areas first, followed by the low concentration zones.

There are potentially thousands of pieces of legislation affecting companies, with more than 100,000 legal requirements relevant to multinational companies (Baker and Sjoberg 2018). Also, laws and regulations tend to differ across industries sectors, markets, and geographies, making compliance more difficult, especially for organizations that have different lines of businesses and multinational organizations that have operations in multiple countries.

In addition, even organizations localized in one country but having business relations with entities from other countries may need to comply with multiple laws and regulations from other countries. Examples of this include the European Union (EU)

Data Protection Initiative (Directive 95/46/EC) and General Data Protection Regulation (GDPR), which call for protecting the privacy of all personal data collected for or about EU citizens even though the organization may not be operating in the European Union. With the advent of the internet and social media, data privacy and protection laws have evolved and have taken a whole new dimension.

"There are 195 countries worldwide, and each may have their own laws and regulations—it is a complex task to be up to date with every country's laws," says Nigel Hawthorn, chief European spokesperson at Skyhigh Networks in relation to data privacy and data protection laws (Carter 2016).

Robert Stroud, Director on ISACA's board and Principal Analyst, Forrester Research says, "Rules and regulations vary widely geographically. There are no consistent guidelines and rules…even neighboring legislative regions have different policies" (Carter 2016).

With globalization, businesses either have offices in other countries or transactions, customers, or suppliers in other countries around the world. It is important to understand the laws and regulatory mandates in other countries to analyze the requirements and understand the implications for multinational organizations as well as organizations conducting business with entities in other countries. It is interesting to note that some countries have multiple regulations and privacy mandates for different types of data. For example, in the United States, there is no single law, decree, or regulation overseeing the collection and processing of personal data. Instead, data protection is regulated by many state and federal laws. For example, the Privacy Act 1974, the Privacy Protection Act 1980, the Gramm-Leach-Bliley Act 1999, the Health Insurance Portability and Accountability Act 1996, and the Fair Credit Reporting Act 2018 (i-Sight 2018).

The complexity of compliance laws, a large number of complex regulatory documents, some of which may be in foreign languages and lack of clarity in the steps required to achieve compliance, strict timeframes, and delivery schedules set by regulators, makes the task of achieving compliance even more challenging.

According to a June 2017 poll by Spiceworks, 37% of respondents listed their top concern with the GDPR as a lack of clarity regarding the steps necessary for GDPR compliance. One way to deal with the problem is by starting on the parts that have more clarity while liaising with regulators to seek clarifications on the less lucid areas (BitSight Tech 2018) to understand the complete meaning and intent of the regulations.

Also, the workflows for data security and privacy are not adequate to meet compliance requirements of recent regulations like GDPR. Revising these workflows, would require analyzing the gaps and then revising policies.

The data requirements related to legislations should not be addressed piecemeal. A holistic approach to data governance that spans across all regulations is needed.

Organizations need to understand the data requirements, impacts of each legislation and regulation, and then find the common or overlapping elements as well as the conflicts and differences between them. For example, two regulations might need reporting of customer account balance, but while one might need it at a product level, the other might require it at an account level. To have a common understanding and consensus, working committees need to be established such that these committees have adequate representations from different business units such as risk, compliance, legal, the business units that use the data, IT, and data management teams that manage the applications, systems, and data repositories.

The working committees need to have a common understanding of the data, how the data is used, the business purposes served by the data, the associated risks, the impact the legislations have on the data, whether the legislations have same or similar requirements from the data (or conflicting requirements), and how the regulations converge on (or conflict the business) usage of data.

A common example of compliance requirement converging with business need, is where both businesses and compliance need more accurate, complete, and consistent data.

Here is an example of a compliance requirement conflicting with a business need. Businesses need access to key data quickly to make decisions, but compliance requires the same data to be secured, access to be controlled, and granted subject to certain checks, such as whether the employee is on an extended leave, his/her job roles, and responsibilities. The additional process overhead, effort, and time needed to gain access to data can be seen as a potential blocker as well as a productivity disruptor. The organizations need to strike a balance between adequately meeting compliance requirements, and at the same time, delivering business value.

Another example of a conflict between a business need and a regulatory requirement is in the area of data retention. Specific business needs may call for longer retention periods than regulatory requirements may allow. For a multinational organization, requirements in one country may not be compatible with another country, and even within countries, regulations in one domain can contradict those in another domain. A collaborative and integrated approach leads to the establishment of policies, processes, roles, and responsibilities around all aspects of data.

In the case of a healthcare company, for example, data subjects who file complaints are sometimes also patients. Based on their privacy-based 'right to be forgotten,' they can demand that such complaints are erased, whereas medical authorities usually demand that such complaints are retained for many years (KPMG Advisory 2018). The key to resolving these conflicts is to establish a working group or council consisting of experts with the relevant backgrounds (for example privacy, regulatory, risk management, compliance, legal, information security, and business processes) which can meet regularly and come up with a suitable data retention solution. A holistic approach to data governance enables an organization to adapt to an increasingly regulated environment, yet remains seamless and transparent across its operations and business systems (Baker and Sjoberg 2018). In cases where data requirements for one legislation is a subset of data requirements of another legislation, the devised data policies, processes, and rules to achieve compliance may apply to the superset of the data.

With compliance-related data sitting in silos across the organizations, there is often confusion as in who in an organization is responsible for the data. In the absence of formal governance, ownership and accountabilities are either not defined or not clearly defined, resulting in "It is not me, it is you" syndrome when such compliance-related issues surface.

Last but not least, the huge costs associated with ensuring compliance is a natural deterrent in the path of implementing programs to achieve compliance.

5.4 Compliance, Data Characteristics, and Activities

Regulations have a direct and indirect relationship with data characteristics and data-related activities along one or more of the following lines (summarized in Fig. 5.5).

Fig. 5.5 Compliance and data characteristics and activities

5.4.1 Data Privacy

Data privacy or information privacy relates to the ability of an organization or individual (to whom the data belongs) to determine what data can be shared with third parties and relates to the confidentiality requirements of the data.

While there is no universally agreed and accepted definition of "information privacy," the following definitions represent major perspectives.

Alan Westin, Professor of Public Law & Government Emeritus, whose research is extensively seen as the first significant work on the problem of consumer data privacy and data protection, defines privacy as:

> The claim of individuals, groups, or institutions to determine for themselves when, how, and to what extent information about them is communicated to others (Westin 1967).

The International Association of Privacy Professionals (IAPP 2019) defines privacy as:

> The appropriate use of personal information under the circumstances. What is appropriate will depend on context, law, and the individual's expectations; also, the right of an individual to control the collection, use, and disclosure of personal information.

Technology alone cannot ensure data privacy, and there are additional factors like individuals and trust that come into play. Data privacy usually comes into play with personal information. Data privacy can restrict data sharing, which can be seen as a competitive disadvantage to deriving business value from data. Anonymization of personal information, while allowing governed access to secured data sets for analytics and tokenized matching can be one solution to this problem (Data Republic 2018).

5.4.2 Data Security

Data security is the process of protecting data from any kind of action that can compromise the data such as unauthorized access, corruption, theft, and distribution. Hence, confidentiality, integrity, accessibility, and availability are the primary security requirements. Data can be protected using technological processes like:

- Encryption,
- Masking,
- Pseudonymization,
- Firewall management, and
- Identity access management, role based accessed and access control lists (ACLs).

The term data protection is often used interchangeably with data security. It can be considered the same as data security, in the context of protecting the data from being accessed by unauthorized users. On the other hand, data protection may also be used for protecting the data from the viewpoint of authorized users, so that this data can be accessed easily later (Difference Between 2019). For example, storing data backups remotely ensure that clean data can be recovered, and is available for authorized users, in case the data is stolen or compromised.

5.4.2.1 Data Privacy Versus Data Security

Data privacy is often confused with data security. Data privacy and data security are very closely interconnected, to the extent that they are often thought of as being synonymous. Data security is more of a technical issue whereas data privacy is a legal one. However, since technology plays a significant role in the collection, exchange, and access of information, data privacy is also perceived as a technical issue.

While data can be protected using technological processes like encryption and masking, the protected data still may not be suitably private. This is because the technology that helps protect data alone cannot ensure data privacy. There are additional factors at play like the individuals who are entrusted with data, and the level of trust that will ensure data privacy.

How information is collected, stored, used, and managed can influence data privacy. Data security helps achieve data privacy. Unprotected data leaves data vulnerable to attacks and thefts. Hence, without data security that can be achieved through technology, it is not possible to have data privacy, as data can be stolen or misused, thus violating data privacy.

Generally, data privacy concerns revolve around personally identifiable information. Most countries and regions have created national data privacy and protection laws. Some of them are GDPR, Canada's Digital Privacy Act, Asia–Pacific Economic Co-Operation (APEC) Privacy Framework, Information Technology Act (India), Federal Data Protection Act 2017 (Germany), the APPI Amendment 2017 (Japan), Privacy Act 1974 (USA), the Privacy Protection Act 1980 (USA), the Gramm-Leach-Bliley Act 1999 (USA), the Health Insurance Portability and Accountability Act 1996 (USA), and the Fair Credit Reporting Act 2018 (USA).

5.4.3 Data Accessibility

Data security and data privacy call for more stringent rules around accessing data. The key things to keep in mind are –

- Who should have access to the data, and
- What type of access should the individual have (for example, read only or read-write access).

Only authorized users should be allowed access to the data. Identity access management tools can be used to authenticate the identity of users and provide role-based access.

5.4.4 Data Completeness

Missing or incomplete compliance-related data elements or records can be a compliance risk. If there are any missing or incomplete data or documentation with regards to regulations that an organization needs to be compliant with, then it is not compliant. For example, Know Your Customer (KYC) laws require an organization to validate the identity of a customer. However, if vital customer information like the address or the date of birth is missing, then it will be deemed as non-compliant.

5.4.5 Data Accuracy

If an organization's critical data is notaccurate, then it cannot adequately manage its risks or make sound decisions. It will be at a competitive disadvantage and will be heading for failure. Fraud prevention, risk aggregation and mitigation, regulatory reporting, better patient care (applicable to the healthcare industry), and responsible lending (applicable to the banking industry) call for the capture and maintenance of accurate data.

5.4.6 Data Consistency

Data in organizations typically reside in silos in many systems resulting in inconsistent data, metadata, and differences in business and data element terminologies. This is because different departments and lines of businesses have created their own systems and data repositories for their specific departmental needs.

These repositories typically have different refresh schedules, resulting in data inconsistencies due to latency issues. However, regulatory reporting requires data to be consistent and requires a single source of truth. Hence, business rules need to be established to define the consolidation of data from these different data sources. Also, data standards need to be established to ensure consistency of data formats and metadata across all data sources in an organization.

Also, if the same data item is being reported in different regulatory reports, the values for the same data item across these reports need to be consistent. For example, there is data overlap between Common Reporting Framework (COREP) and Financial Reporting Framework (FINREP), which are regulatory reporting frameworks required by the European Banking Authority. These overlapping data item values need to be consistent across both the reporting frameworks and each of the reporting submissions.

5.4.7 Data Integrity

Data integrity concerning compliance with regulations is maintaining and guaranteeing the accuracy, completeness, and consistency of data over its entire lifecycle, each of which we have discussed earlier in this section.

Data integrity was a hot topic of discussion in the pharmaceutical industry throughout 2016 when non-compliances made headlines and data integrity issues triggered more than one-third of all regulatory actions. While Indian and Chinese companies dominated 'non-compliance news', 2016 proved that data integrity concerns were not limited to these two countries. European operations of major drug companies like GlaxoSmithKline, Teva and Otsuka as well as US firms were also found to have serious non-compliance concerns (PharmaCompass 2017).

Concerning current good manufacturing practices (cGMP) regulation for drugs, data integrity refers to the completeness, consistency, and accuracy of data. Complete, consistent, and accurate data should be attributable, legible, contemporaneously recorded, original or a true copy, and accurate (ALCOA).

Data integrity is critical throughout the cGMP data life cycle, including the creation/generation, transformation, processing, maintenance, archival, retrieval, transmission, and purging of data post the retention period of the record. System design and controls should enable easy detection of errors, omissions, and aberrant results throughout the data's life cycle (FDA et al. 2018; McDowall 2018).

5.4.8 Data Currency and Timeliness

Timely, as well as current data, is extremely important for regulators. Compliance with regulations requires organizations to generate up-to-date data at the required level of granularity in a timely manner. Timestamps can help determine the currency of data, and can be used to validate currency for regulatory purposes.

5.4.9 Traceability

To comply with regulations, organizations need to understand the lifecycle of their data (Franco 2018) and must demonstrate the same to regulators as per a pre-established schedule if one exists, or on a demand basis.

Hence, organizations must be cognizant of the data flows and be able to trace the data to its originating source and understand the processes, transformations, and the trajectories of systems that the data has gone through over time. Organizations should know what platforms house data related to business functions. Data may be created in one system but is provisioned to the downstream systems by another system. Without knowing where the data item originated from and its tranformation as it moved across systems, it is not possible to guarantee and prove the quality of the data from a regulatory standpoint. In simple words, traceability answers the following questions:

- What is the source of this data? That is, all the upstream source systems as well as the originating source.
- What is the target for this data? That is, all the downstream target systems.

5.4.10 Data Granularity

As new scandals surface, regulations evolve, and these regulations require more granular data. This is because the detailed data can reveal data flaws that can be missed in aggregate data. The saying "the devil is in the detail" applies here.

Organizations capture and store data in different systems at different levels of granularity. This can be challenging when consolidating and aggregating data elements required for regulatory reporting from these different systems. For example, the European Banking Authority (EBA) has two reporting frameworks: the Common Reporting Framework (COREP) which contains 28,486 data points across 31 reports, and the Financial Reporting Framework (FINREP) which contains 6,682 data points across 32 reports (Source: EBA). These reporting frameworks require data at a much more granular level.

5.4.11 Data Usage

Data usage is controlled or limited by external and internal regulations, laws, policies, and data usage to derive business value needs to be compliant with regulations. It is important to understand which activities or actions are permitted and which are prohibited to be compliant. These are generally defined by the data usage policies to ensure that the data is not misused or abused. Data classification that helps segregation of data elements by their sensitivity and risk quotient can help understand and define the data usage requirements.

5.4.12 Data Retention

Compliance with regulations has an impact on data retention. Data archiving and purging helps organizations remove data that is no longer needed for business operations but could pose security and/or privacy risks if exposed or stolen.

Organizations need to establish their own data retention requirements by taking into account the data retention laws as well as the data domain or subject area, data entities, and data elements that are impacted by these laws.

For example, healthcare organizations are subject to the Health Insurance Portability and Accountability Act (HIPAA) data retention requirements. The organizations that accept credit cards must adhere to a Payment Card Industry Data Security Standard (PCI DSS) data retention and disposal policy. The General Data Protection Regulation (GDPR) applies to the personal data related to EU citizens, whether or not the organization collecting the data is in the EU, as well as any individuals and organizations whose data is stored within the European Union. It's critical to have a data retention policy that explains which data is being held, why and where it's being held, and for how long, as it relates to GDPR directives (Rouse).

5.5 Data Characteristics and Compliance—Interrelationships

Some of the above data characteristics and activities that have been discussed in the previous section are interrelated. For example, data accuracy is impacted by data completeness, timeliness, and data currency. Data privacy and data security have an inverse relationship with data accessibility and data usage. In other words, high privacy and security requirements can restrict data access and usage. Also, data privacy might drive specific data retention requirements.

While data protection and privacy laws have a direct relationship with data privacy, data security, data accessibility, data traceability, data usage, and data retention aspects, they have an indirect relationship with data accuracy, data currency, timeliness, completeness, and consistency.

For example, the US Health Insurance Portability and Accountability Act of 1996 (HIPAA), creates national standards to protect individuals' medical records and other personal health information by setting boundaries on the use and release of personal health information and requiring health providers to establish suitable safeguards to protect privacy of individuals' health information.

This means that an individual's health information should not fall into the wrong hands. When medical records are sent to the patients, if the patient's address is not up to date or is not accurately captured or is incomplete, the records will not be received by the intended individual but fall into someone else's hands, who should not see it thus resulting in a HIPAA violation.

While regulatory reporting has a direct relationship with data accuracy, data completeness, data consistency, data timeliness, data traceability, data accuracy, other aspects like data security, data accessibility, data retention, and data usage would also need to be taken into consideration as you would still need to protect the data from being accidentally or unlawfully modified or deleted by unauthorized users.

For your compliance program to work, you need good quality data, robust processes, efficient systems, and an overarching data governance framework to ensure that good quality data are captured, managed, used, archived, and disposed in line with compliance requirements. This is because poor quality data, inefficient processes, and systems can result in flagging false violations or not detecting actual violations, and thus, not raising the red flags, resulting in additional work for the compliance team. Also, inefficient processes and systems create additional unnecessary data, which is a burden from a cost, storage, and management perspective.

5.6 How Does Data Governance Help with Being Compliant?

Regulations and data have an impact on each other, and to be compliant with the data-related requirements of regulations, organizations need an effective data governance program in place.

Achieving data-related requirements of regulations generally consists of the following two parts (illustrated in Fig. 5.6):

- the implementation of a technical solution to achieve compliance and establishing a data governance program to provide the framework to support the compliance related tasks, and
- answer the "five Ws and one H" in relation to compliance.

Data governance helps organizations in their data compliance journeys in the following ways (summarized in Fig. 5.7):

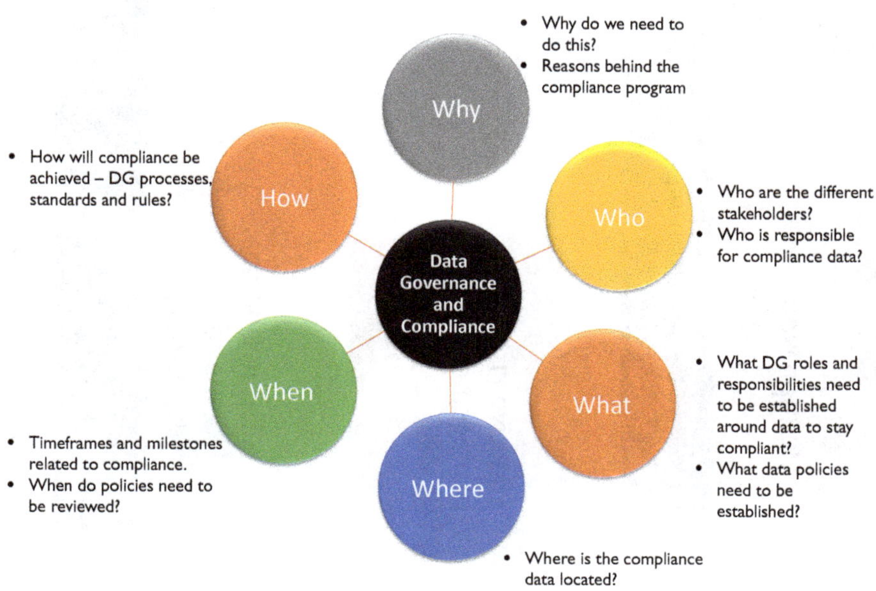

Fig. 5.6 Data governance and compliance—the five Ws and one H

Fig. 5.7 Role of data governance in compliance

5.6.1 Improved Data Discovery And Data Lineage

Compliance requirements require the ability to locate and understand data associated with regulations. Data lineage or traceability, the ability to trace data present in a report to its sources, is becoming a regulatory requirement, especially in cases of regulatory reporting.

The reporting entities must be able to demonstrate how each value in a report is generated, including its calculation, transformation details, and source data. Each data item in a regulatory report must be credited to an owner and a steward, who should be able to prove that it is of high quality. Also, organizations should have the knowledge and capability to trace data back to an authoritative source.

Knowing where your compliance-related data resides is the first step towards achieving compliance and is often the biggest challenge too. This is because compliance-related data is generally stored in a myriad of different locations, systems, data stores, and formats. In short, compliance-related data is often scattered.

However, unless you know where the related data is stored and can locate the data in the vast ocean of data, how can you produce it in regulatory reports or adequately protect it or prevent unauthorized access or ensure its privacy or guarantee if it is fit for purpose?

Data governance results in data stewardship roles established. Data stewards who are SMEs are responsible for data sets, understand the business meaning of the data, and the purposes for which the data is used. They can locate the data needed for compliance purposes with the help of data discovery software.

Also, data governance results in the establishment of standards around data elements and data entities. Data governance ensures that metadata and data lineage information is up-to-date, which in turn assists the data discovery process.

The data stewards from different business units/ departments/ business functions liaise with each other to sort out the differences related to the data. Data standards and good quality metadata help correct understanding of data elements and help understand the data overlaps between multiple systems.

5.6.2 Enhanced Data Privacy and Security

Data should be maintained securely throughout its lifecycle. It should be accessed and managed only by authorized individuals. Data governance ensures that adequate policies, guidelines, processes, and controls are in place relating to access, usage, storage, and sharing of critical data. This is particularly important with the personal data that is used in anti-money laundering (AML) know your customer (KYC) checks (Pappu 2018) and to meet the data protection and privacy laws such as GDPR.

5.6.3 *Improved Data Quality*

Data quality is at the core of compliance. Every compliance program has some data elements mapped that need to be of high quality. Quality data is key when it comes to satisfying and complying with the rigorous controls generally imposed by regulations, especially the ones involving regulatory reporting requirements. Data governance results in better quality data, that is, data that is complete, current, accurate, consistent, and free of duplicates, by having appropriate controls in place to monitor the same.

5.6.4 *Effective Risk Management*

Effective risk management is necessary to ensure compliance. Effective data governance enables the identification, monitoring, management of risks throughout the organization, facilitates organizations to take quick corrective action where required, and lessens the likelihood and severity of damages.

5.6.5 *Accountabilities and Responsibilities*

Regulations require executives to be aware of as well as be responsible for the data that drives their businesses. Certain privacy and protection laws require an appointment of a Data Protection Officer (DPO) or a Chief Privacy Officer (CPO).

The accountability and ownership of data are critical when it comes to compliance-related tasks. A clear understanding and agreement amongst all stakeholders as to who is responsible for what in the organization concerning compliance-related tasks, and ensuring that they have the right authorities to carry out these tasks and are sufficiently empowered, is important to achieve compliance.

With data residing in silos and with multiple consumers of data, achieving data compliance is certainly not a "one-man show," and it often requires the involvement of multiple stakeholders. Clear responsibilities need to be established around the maintenance, approvals around the data, proper communication channels need to established, and a collaborative environment needs to be fostered to avoid different teams duplicating the same tasks or being in conflict when it comes to achieving their respective compliance objectives.

Data governance results in establishment of roles, responsibilities, ownership, stewardship, accountabilities, and decision rights in case of a conflict resolution. Data governance also ensures that appropriate structures, roles, and working committees are there in place to establish, enforce, maintain data related policies, processes, standards, rules, and metrics to ensure compliance with regulations. Being able to exhibit effective ownership and stewardship of data items and data sets is crucial to both regulatory compliance and data governance.

5.6.6 Policies and Processes

In the words of Shannon Fuller, Director of Governance Advisory Services, Gray Matter Analytics, "Regulations help to shape data governance policy and the implementation of those policies reduce the risk of violating those regulations."

An organization should have internal data policies, which should align with the external regulatory requirements, should be up-to-date, and reflect the most current regulatory requirements. Hence, there should be checks and controls in place to review the external regulatory documentation periodically, look out for and understand changes if any, and revise internal data policies accordingly to stay aligned with the external regulatory requirements.

Depending on the external regulatory requirements, one or more of the following data policies would need to be aligned with the external regulations: data classification policy, data capture policy, data security and data privacy policies, data usage policy, data access policy, data retention policy, data governance structure policy, and data quality policy. Data-related processes, standards, and rules would need to be modified, or new processes, standards, and rules would need to be created to adhere to data policies so that the data related processes are compliant with the regulations.

5.6.7 Metrics

An effective data governance policy results in the creation and monitoring of data governance metrics. Tracking metrics such as number of security incidents, data quality metrics for compliance-related data (such as percentage data accuracy, percentage data completeness, and percentage data duplication), percentage of data elements that cannot be traced back to the source, number of failed data audits, and turnaround time for audit queries can provide insights into the health of your compliance program and corrective actions can be planned and implemented if needed.

5.6.8 Education, Training, and Change Management

Effective data governance ensures that stakeholders are educated and trained in the data governance approach that is used to achieve compliance. It also ensures that change management is in place to ensure that stakeholders let go of their old way of working and get them to feel comfortable doing things with the new approach. In the absence of awareness, training sessions, and effective change management, below sentiments and attitudes are common:

> We have always done things this way and have not got into trouble.
>
> Everyone else does it this way.

Let us just correct the problem moving forward.

We have other high priority tasks.

However, if the staff maintain the above attitude towards compliance, then they will not do things the right way. In relation to data, compliance requirements cross departmental boundaries, and in the case of multinational organizations, they cross country boundaries too. There are multiple stakeholders concerning data, and to ensure that compliance is achieved, all of these stakeholders need to be educated and trained in the use of data governance policies, processes, standards, and controls so that they understand their responsibilities towards data and learn how to manage data differently to be compliant. They should also understand why they are required to do things differently and must adopt the ability to distinguish between what is right and wrong. Change management ensures that staff embraces new ways of working and do not revert to old ways.

Not all stakeholders have the same roles and responsibilities involving compliance. Thus, training, communication, and awareness sessions need to be tailored to the different stakeholder groups based on the level of involvement required from them, and their data-related roles and responsibilities concerning compliance.

Table 5.1 shows some important regulations and how data governance can help achieve compliance with these regulations.

5.7 Regulatory Compliance and Data Management

Regulatory documents, for example, GPDR, HIPAA, and FATCA, are long and complex documents, with sections and subsections that need to be understood well to assess their implications on the organization's business processes and data.

It is important to understand how people, data, metadata, business functions, and processes map to the regulatory mandates. It is also necessary to define internal rules that comply with the rules specified in the external regulations. When we talk about data in the context of compliance, it refers to data residing digitally and data in physical documents as well as data in rest and data in motion. The questions to ask are:

- What data is needed for compliance purposes (for example, customer data, address data, transaction data, and finance data)?
- What business processes generate and use these data?
- What specific data elements are needed and are available for compliance purposes?
- What data elements are not captured (that is, they are not available), but are still needed for compliance purposes?
- What are the data quality dimensions that need to be considered for compliance purposes?
- What is an acceptable level of data quality to meet compliance? (This is specifically important when reporting compliance data or when adhering to certain reporting standards.)

Table 5.1 Some important regulations and the role of data governance

Regulations	Brief description	How can data governance help?
Health Insurance Portability and Accountability Act (HIPAA)	The Health Insurance Portability and Accountability Act is a US regulation that sets the standard for protecting the confidentiality of sensitive patient data and protected health information stored both electronically and physically (TOC_HIT). Chap. 4 discusses HIPAA in greater detail	Ensures adequate policies, processes, standards, and controls are there in place for capture, storage, classification, access, usage, transmission, sharing, retention, security, quality of protected health information of individuals; sensitive patient data, issue/incident management; facilitates building metrics; and audit mechanisms are in place to ensure compliance with these policies; Helps discovery and classification of in scope patient and health data; Helps standardization of definitions, names, and metadata related to medical terminology such as health symptoms, diseases, procedures, and medications; Ensures that training programs are in place to educate employees about data-related risks and requirements; Ensures that clearly-defined governance roles, responsibilities, accountabilities, and working groups are established to ensure that in-scope data is adequately captured, protected, shared, and transmitted, and that any data breaches or incidents are reported in line with the violation requirements.

(continued)

Table 5.1 (continued)

Regulations	Brief description	How can data governance help?
Anti-Money Laundering and Counter-Terrorism Financing Act (AML/CTF Act) and Know Your Customer	Anti-Money Laundering and Counter-Terrorism Financing Act is an Australian regulation that regulates bullion, gambling, financial, and remittance services, where the conversion and transfer of physical and electronic forms of money are vulnerable to money laundering and terrorism financing KYC, is the process of a business verifying the identity of its clients and assessing their suitability, along with the potential risks of illegal intentions towards the business relationship. The term is also used to refer to the bank regulations and anti-money laundering regulations which govern these activities [Wikipedia]. Chap. 4 discusses AML and KYC in greater detail.	To be compliant, business processes require high-quality customer data for customer risk scoring and financial transaction monitoring to clearly identify customers within high-value transactions. The customer account data and financial transactional data required for compliance typically resides in a number of different sources. It needs to be consolidated to provide a single view. Data governance ensures that adequate policies, and processes related to risk classification of data, data usage, data access, conflict resolution, and data quality are in place. Data governance helps standardize customer data across the organization. Data governance helps establish clear lines of responsibility among the business process owners, their technology counterparts, and ensures that fraud and financial crime management teams are in place to improve data quality and ensure correct usage of data. Data governance ensures that adequate metrics, controls, monitoring processes, and mechanisms are in place to measure the quality of customer and financial transaction data.

(continued)

Table 5.1 (continued)

Regulations	Brief description	How can data governance help?
Regulatory Reporting (example– FINREP, COREP required by European Banking Authority, Common Reporting Standard for Tax)	The Financial Reporting Standard (FinREP) framework governs the reporting of financial information annually. Common Reporting Standard (CoREP) framework governs the reporting of risk (credit, market, operational, solvency, and capital) monthly and quarterly. Chap. 4 discusses FinREP and CoREP in greater detail. AnaCredit contains granular information about individual bank credit exposures.	Regulatory reporting requires reporting of a large number of attributes and various slices and dices of data, with reporting data elements that might be derived through calculations or are sourced from different data sources. Data governance ensures standardization of data across departments, brands, and groups, including the use of common definitions and common formats. Data governance also ensures that the data can be traced back to the sources, and data stewards are established who know the rules involved in the calculation of the reporting elements. This ensures that mechanisms and metrics are there in place to monitor data quality and raise flags when quality falls below the threshold defined in the data quality policies. Regulatory reporting often results in the creation of a compliance data mart within a data warehouse or new warehouses, if the present data warehouse model does not cater to the data granularity levels required to meet regulatory requirements. Data governance ensures that governance structures (including working committee, data ownership, and data stewardship), policies and processes are in place to adequately source and-integrate data from the data silos, to audit and attest reporting data, resolve data conflicts, guide role-based access, and protect reporting data from unauthorized access and manipulation.

(continued)

Table 5.1 (continued)

Regulations	Brief description	How can data governance help?
GDPR	GDPR is a set of data protection laws that cover the management, use, and security of personal data and applies globally to organizations that process personal data about European Union (EU) data subjects with an intent to better protect personal data; EU data subject simply means any EU customer, consumer, partner, staff, or other individual (ICO 2019). Chap. 4 discusses GDPR in greater detail.	Data governance ensures the establishment of policies and processes for: • data discovery, • data classification and defining categories of in scope personal data sensitivity, • protection of the in-scope personal data, • retention of in scope personal data, • access of in scope personal data, • transmission and sharing of personal data, • timely report of a breach, • fix personal data quality issues, • subject access request, • getting consent from individuals, • right to be forgotten, and • data portability. Data governance facilitates discovery and traceability of in scope personal data, and prioritization of data sets containing in-scope personal data. Data governance ensures the creation of controls or enhancement of existing controls to protect data sets containing in scope sensitive data, and mechanisms and metrics to monitor and measure access, anomalous user behavior, the usage of data, and data transmissions across organization borders. Data governance ensures ownership and stewardship of in-scope personal data and creation of governance structures, including working groups and committees. Their roles and responsibilities are to define policies, processes, controls, and standards to manage in scope personal data in line with GDPR requirements and governance of the single view of personal data. For a single view of personal data, there should be adequate conflict resolution processes to resolve conflicts originating from personal data residing in silos in multiple disparate data sources.

(continued)

Table 5.1 (continued)

Regulations	Brief description	How can data governance help?
Basel Committee on Banking Supervision 's standard number 239 (BCBS-239)	BCBS 239 focuses on risk data aggregation (RDA), reporting, data quality, and lineage to improve risk management. Chap. 4 discusses BCBS-239 in greater detail.	BCBS 239 specifically asks for data governance and requires risk data aggregation capabilities and risk reporting practices to be subject to strong governance and high standards of validation (BCBS 2013). Data governance helps in achieving uniform naming conventions across the banking group and establishing roles and responsibilities about ownership, stewardship, and quality of risk data, which are also key requirements of BCBS 239 Data governance ensures that policies, processes, and standards are in place to produce complete, accurate, and reliable risk data promptly in both normal and stress/crisis scenarios, for appropriate usage, and access of data as well as for resolving data quality issues and conflicts. Data governance also ensures that metrics and monitoring mechanisms are in place to measure the quality of the risk data.
Payment Card Industry Data Security Standard (PCI DSS 2010)	PCI DSS is a set of standards created to help protect customers' credit card and debit card information (PCI SSC 2010; Wiki-PCI-DSS). Chap. 4 discusses PCI DSS in greater detail.	Data governance ensures that appropriate policies and processes are in place for protection, access, retention, usage, transmission of cardholder data, and response to an incident. Data governance ensures that organization structures, roles, responsibilities, ownership, and stewardship are established for governance of cardholder data. Data governance ensures that controls, metrics, and monitoring mechanisms are in place to monitor data transmission, access, and usage of cardholder data.

(continued)

Table 5.1 (continued)

Regulations	Brief description	How can data governance help?
Sarbanes–Oxley Act (SoX)	SOX was designed to implement corporate accounting and financial disclosure requirements that increase transparency in corporate governance, and financial reporting, and formalizes a system of internal checks and balances and applicable to all publicly held American companies, any international companies that have registered equity or debt securities with the U.S. Securities and Exchange Commission (SEC) and accounting firm or another third party that provides financial services to either of these. Chap. 1 discusses SoX in greater detail.	Corporate accounting and financial reporting are associated with the creation, storage, transmission, and maintenance of financial data. Data governance concerning SoX would be related to establishing adequate policies, processes, roles and responsibilities, and controls to govern this data in relation to lineage, usage, access, sharing, transmission, protection from manipulation, and ensuring high quality of the related data that is free from unauthorized manipulation. The financial data required for reporting for SoX compliance typically reside in a multitude of systems. The data standards and consolidation rules need to be defined to build a target system containing the consolidated data required for compliance reporting.
FATCA (foreign account tax compliance act)	FATCA (foreign account tax compliance act) is a US law that was designed to ensure that US citizens who hold assets outside the US and offshore accounts, pay relevant taxes Chap. 4 discusses FATCA in greater detail.	FATCA compliance would require foreign financial institutions to identify whether client data belongs to a US taxpayer. Typically, organizations have multiple clients on-boarding systems. To adequately analyze and report on the required data elements, they would need to consolidate data from the different systems and store compliance-related data in a compliance data store. Data governance would involve creating data governance organizations structures such as working groups, and steering committees with the impacted stakeholders, establishing roles and responsibilities, policies, and processes to discover and classify the related data and ensure compliance effectively, and ensure standards, and rules are in place to consolidate, analyze the data, and metrics to ensure that the data is fit for purposes.

(continued)

Table 5.1 (continued)

Regulations	Brief description	How can data governance help?
Dodd-Frank Wall Street Reform and Consumer Protection Act commonly known as Dodd-Frank Act	Dodd-Frank Act was introduced in the aftermath of the 2008 financial crisis to ensure transparency in record keeping and prevent a repeat of the financial meltdowns. Chap. 1 discusses Dodd-Frank Act in greater detail.	Dodd-Frank Act is focused on ensuring the transparency of instrument reference data or contractual information related to financial "instruments" like equities, bonds and derivatives; entity reference data, which is used for counterparty risk assessment and defines who is doing business with whom; pricing data; and information about portfolio holdings (Brunelli; Guess 2011) which would require this information to be of high quality. Data governance ensures that these data are standardized so that the data is consistent. Data governance also ensures that the roles and responsibilities, policies, processes, and controls are in place to ensure that the data is of high quality and are audited at regular intervals.

- Who is accountable and responsible for ensuring that the data delivered is of acceptable quality?
- What data elements or data sets should be protected through masking or encryption?
- What data sets should be transferred securely?
- Where do your data repositories reside?
- Who controls access to data?
- Who has access to what data?
- What physical and logical protections are in place to keep your data repositories safe?

It is very important to have a common understanding and consensus amongst stakeholders about the data requirements required to meet regulatory compliance. The data policies should align with the regulatory requirements, with all stakeholders having a common understanding of the policies.

Data management for compliance involves the following activities with data governance as an overseer of these activities:

- Define and discover data,
- Data profiling and assessment,
- Compliance data store—data mapping and transformation,
- Logging, and
- Data standardization.

5.7.1 *Define and Discover Data*

You first need to understand and define the data required for compliance. You need to map the business data elements to the actual data elements stored in the various database systems and files in the organization. For this, you need to have an understanding of the data sources, the data flows, the intermediate systems, and the destinations of data.

For example, consider Personally Identifiable Information. Personally Identifiable Information (PII) includes:

(1) any information that can be used to distinguish or trace an individual's identity, such as name, Social Security Number (SSN), date of birth, place of birth, mother's maiden name, and/or biometric records; (OMB Memorandum M-07–16 2007) and,

(2) any other information that is linked or linkable to an individual, such as medical, educational, financial, and employment information (OMB Memorandum M-07–16 2007).

An individual's name might map to different data elements, such as first name, middle name, last name, alias, or the full name depending upon how the information related to an individual's name is stored in an database.

Also, since an individual's name is usually stored in multiple tables and in multiple disparate data stores with data elements holding the information named differently, all these data elements need to be located despite the differences in naming conventions and the data flow also need to be discovered.

While some data element names might closely relate to the real-world entity attribute that it represents, others might not be that revealing. For example, any of the following data element names indicate the Social Security Number:

- SSN_Num
- Social_Security_Number
- Social_Security_No
- Social_Security

However, Social Security Number may be stored in data elements that do not by themselves allude to storing Social Security Numbers such as the following and hence, require inputs from data domain and subject matter experts (SMEs) to locate them:

- CUST_NUM
- SSID
- SO_SE_NU

Organizations hoard and store a large amount of data, but compliance-related data comprises a tiny percentage of this total data. Locating compliance-related data is an expensive exercise that needs to be facilitated using data discovery tools.

It is important to classify the data elements depending on the security requirements and as per the organization's data classification policy, to secure this sensitive data. These aspects will be discussed in detail within the second book of this series: Data Governance and Data Management.

5.7.2 Data Profiling and Assessment

From a compliance perspective, data profiling and assessment covers the following aspects:

5.7.2.1 Assessing the Quality of Source Data

Compliance data needs to be accurate, current, timely, complete, and consistent. It is necessary to assess the source data and to ensure that the data sources are authoritative, and trustworthy, and the data is of high quality.

Data governance ensures that policies and standards that define the required data quality levels are in place and are enforced and followed through appropriate processes, control, and metrics.

If the data is not of adequate quality, then root cause analysis needs to be conducted to understand the underlying cause of bad data. Data remediation and improvement solutions need to be designed and implemented depending on the root causes. For example, the underlying causes may be faulty business processes or lack of awareness.

Also, what needs to be assessed is the granularity of the source data. Data granularity is the level of detail represented by the data. Different systems store data at different levels of granularity. High granularity means that you get minute details of the data and transactions. Lower granularity means a more summarized view of the data and transactions. Current regulations have been marked by an increased level of data granularity necessary for compliance reporting.

For example, AnaCredit, which is the abbreviated version of the analytical credit dataset, was adopted by the governing council of the ECB, and is a new dataset containing detailed information about individual bank credit exposures, that is granular credit and credit risk data in the EU area. These data should be collected in a common register called AnaCredit. The saying "devil is in the detail" applies here.

Rather than just requiring aggregated statistics, compliance with many new regulations requires a single and detailed view of entities like customer and underlying transactions. This is reinforced by Tony King, EMEA Financial Services Sales Manager, Violin Memory, who states:

> Regulators, compliance and risk officers are demanding more granularity, based on regular high-level reporting, and often at increased frequency. This means that large and complex data sets need to be drawn from multiple sources, consolidated, aggregated and reported on. Banks that try to take a central data warehouse approach are finding that not only can this be an incredibly complex and expensive undertaking, but that meaningful results are very difficult to achieve in a timely manner, particularly if that data warehouse is disk-based (Hall 2016).

While data remediation is a reactive approach to fixing data issues and treats the symptoms rather than the root cause, it is the only option once data is created, as bad data if discovered by regulators can result in huge fines and may also lead to faulty decision making.

However, depending on the underlying causes, preventive solutions also need to be designed to ensure that the data is captured and processed correctly. These could be process improvement solutions, automated data capture, incorporating validation checks to catch missing values or inappropriate entries, and forcing the user to enter appropriate values, training data entry operators to record data correctly, as well as defining standards and processes around data collection.

Data governance ensures that roles and responsibilities are there in place to take ownership of data issues and resolve them in a timely manner. It also ensures that processes are established to triage, escalate, and resolve issues, and that data collection standards and processes are in place and enforced.

5.7.2.2 Assessing User Access Data

Depending on the data classification and regulatory security requirements around the data, business process requirements, and access to data, segregation of duties should be determined by roles, responsibilities, and job descriptions of the individuals.

The user access list containing the employee details and access to applications need to be assessed periodically to ensure that the right individuals have the right level of access and do not have conflicting access permissions that violate the principle of segregation of duties to reduce the risk of data loss, theft, and fraud.

For example, employees leave organizations, go on extended leave, move from one department to another, or are deputed to work in other job locations. Such individuals do not need access to applications they were earlier entitled to. An analysis of user access data sets helps to discover illicit access and revoke access for users who should not be having access to an application.

Data sets typically contain many data elements, some of which can be accessed by a wider user base while some will have a restricted user base. Robust security controls need to be in place to hide data contained in data elements that need to be hidden from users who should not be able to view this sensitive data while giving users who should have access, easy visibility to the same data.

Data governance ensures that data access, data security, data usage policies are in place and enforced, that roles and responsibilities, and controls have been established to monitor the robustness, validation of the security and access management solutions.

5.7.2.3 Assessing Quality of the Target Compliance Data Store/Data Warehouse/Data Mart

Regulators often want to have a single source of truth or a single view of compliance data, which results in consolidating data from different data sources and storing them in a target compliance data store or data warehouse or data mart within the data warehouse. Data ownership and data stewardship need to be established for the target data system that stores the compliance data, and the data needs to be assessed to make sure that the data set meet the data quality standards and thresholds.

5.7.3 Compliance Data Store—Data Mapping and Transformation

Organizations typically have a complex data landscape with data stored in a variety of formats, at different levels of granularity, and in silos in a large number of disparate systems. This results in data inconsistencies and multiple versions of truth that create doubts around data accuracy and questions regarding the credibility of data, resulting in a loss of confidence in the data.

For regulatory purposes, the data elements required for compliance can be extracted from the source systems. They can be transformed and stored in a compliance datamart in the corporate data warehouse, or compliance data stores, or compliance system to present a single view of the compliance data to the regulators. By creating a dedicated compliance data store, organizations can ensure easy compliance data discovery and analysis. The following must be kept in mind when designing the target compliance data store:

1. The target data elements need to map to the right data element or set of data elements from the source systems.
2. In case source data has quality issues, the data need to be cleansed before storing in the target compliance data store.
3. Right transformation rules need to be devised where necessary, to meet compliance requirements. This could involve formatting requirements or currency conversions, or mathematical calculations involving source data elements. The data elements in organizations are spread across numerous systems. Regulatory reporting requires sourcing data from the different systems that store data in different formats at different levels of granularity. With the same data element residing in multiple systems, lack of consistency, and missing data are common problems. A decision needs to be made as to the priority of systems from which a data element needs to be sourced from, in case of incompleteness and inconsistencies. For example, System A, System B, and System C store customer data. However, System A stores the most current customer data and has data updates once every day, System B has data updates once every week, and System C has data updates once every fortnight.

   Since compliance requires the most accurate and current data, the transformation rule will be to source the data from system A. If customer data was missing in system A, then the data would be sourced from System B, and in case the customer data was missing in System A and B, then the data would be sourced from System C. So, the priority of sourcing data from the systems are in following order:

   1st Priority—source customer data from system A.
   2nd Priority—source customer data from system B.
   3rd Priority—source customer data from system C.

   Additionally, the data might need to be aggregated or transformed or reformatted before storing in the compliance data store.
4. Data governance is necessary to ensure that there is collaboration and consensus amongst stakeholders of all the source systems and the target compliance data store, roles and responsibilities are in place for resolution of data quality issues in the source systems, and that policies, processes, rules, and standards are established, enforced, and revised periodically for capture, transformations, access, security, storage, and retention of compliance related data. From a people point of view, to achieve compliance, you need business, IT, data, risk, legal, and compliance stakeholders to work together. Data governance brings together the stakeholders from these diverse steams to ensure that related data requirements of compliance are achieved.

While having a compliance data store as a single source of truth for regulators provides reliable data that is desirable for compliance purposes and serves well for regulations where data accuracy, consistency, and completeness are the required characteristics; regulations which deal with data security and privacy, would need in scope compliance data scattered in different parts of the organizations to be located, the risks associated with them to be analyzed, priorities established and roles and responsibilities, processes, policies, metrics, controls, and technology solutions to ensure that these data are managed, accessed, and used as per the compliance requirements.

You would like to first identify the high-risk areas that have large amounts of in-scope compliance data as opposed to the low-risk areas that have very few in-scope compliance data. Also, more rigorous data governance practices need to be established for the high-risk areas as opposed to the low-risk data areas.

Even with a single source of truth, regulations might need traceability of data to the original source systems, and you would need appropriate governance to ensure that data lineage is captured and that there are individuals in charge to maintain these. Also, data quality metrics and threshold values need to be established, and data quality needs to be monitored at regular intervals. Figure 5.8 shows the data management activities around establishing a compliance data store.

5.7.4 Logging

Logging is one of the most important means to warranty visibility and obtain insights into the data-based events/actions of a system, network, or IT environment, and is an essential compliance requirement. You should have logs and audit trails to trace actions and to be able to detect what went wrong as well as prove to auditors that you are indeed compliant. Effective data governance will result in the establishment of effective controls, processes, and roles to ensure periodic surveillance is conducted, and risks are discovered and mitigated.

5.7.5 Data Standardization

Data standards provide specifications and rules for the definition and creation of data entities and data elements. Data standards include naming conventions, data types, length, value ranges, domain of values, formats, classification types, metadata standards, as well as national, international, and industry data standards (for example, SIC, ISO, and HL7).

Data standardization improves data quality, reduces data inconsistency, and improves compliance. However, effective data governance is required to achieve and maintain it.

Also, what needs to be taken into consideration is the need for standardization of data across all systems. The standardization efforts should be conducted with due awareness for industry-wide standardization initiatives, and organizations should define standards accordingly. For example, banks' data input standardization efforts are conducted with due awareness for industry-wide standardization initiatives such as the use of legal entity identifiers (LEIs), and message standards frameworks such as ISO 20022 (O'Hara 2018).

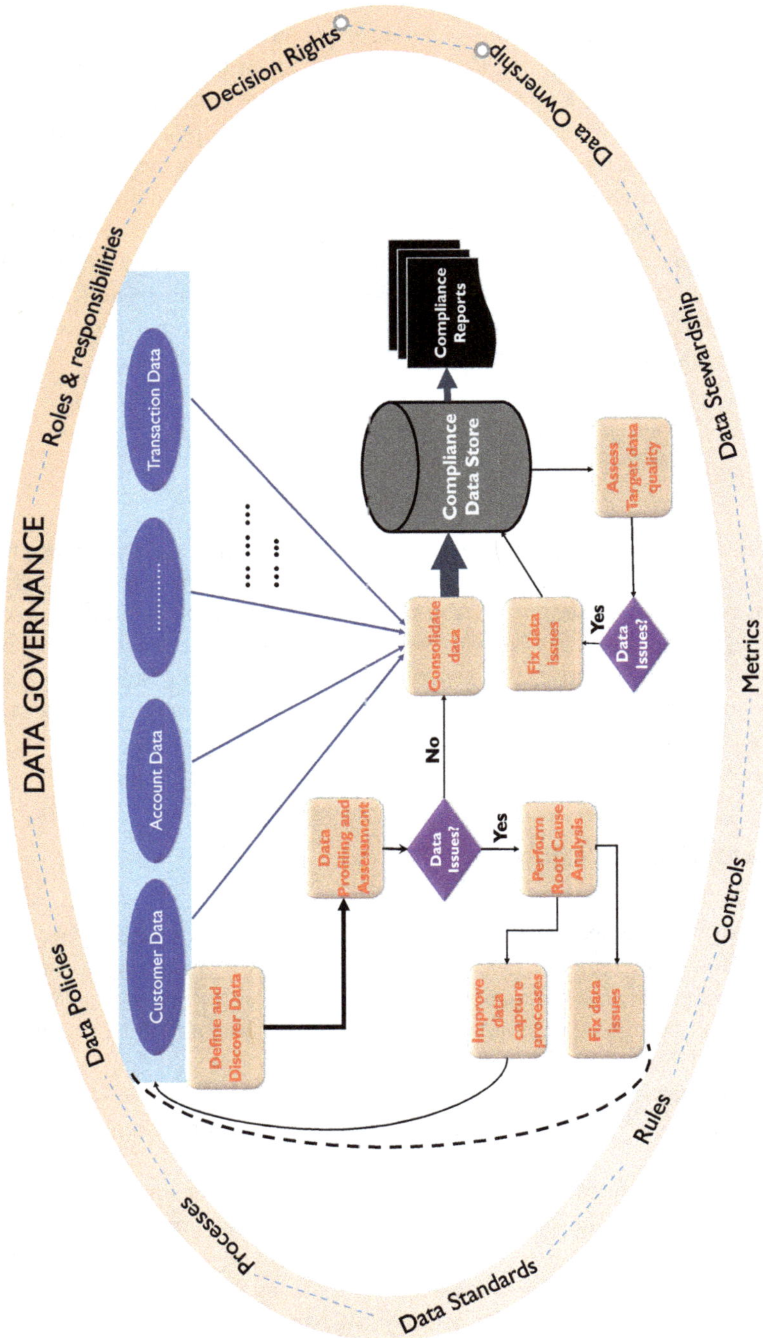

Fig. 5.8 Regulatory compliance and data management

5.8 Concluding Thoughts

Compliance comes in a range of forms and spans different industry sectors, with the finance and healthcare industries being at the top of the list. The regulatory landscape is diversified by industry sectors, markets, and geographies, which are extremely dynamic and constantly evolving.

The number of regulations that require an organization to have control over data has increased over the last few years. It includes well-known mandates such as the GDPR, the International Financial Reporting Standards (IFRS), Health Insurance Portability and Accountability Act (HIPAA), and the Payment Card Industry Data Security Standard (PCI DSS).

There needs to be a balance between complying with regulations and extracting business value from the data that organizations deem critical for their operations and business strategy. Most organizations already have an existing compliance program, and some form of data governance is already used in operations. What is necessary is to understand the gaps in the data governance practices to support effective compliance instead of starting from scratch.

It is important to look at all the compliance regulations holistically to understand overlapping or conflicting requirements and then devise an optimal solution.

For example, GDPR, AML, and FATCA target customer data and would require a single high-quality view of the customer. However, while GDPR only targets EU subjects, FATCA is focused on the US individuals, and AML focuses on all customers. GDPR focuses on privacy, which would restrict sharing of data, while FATCA would require sharing financial asset data of US citizens. If a US citizen is also an EU citizen, then processes would need to be defined by participating financial institutions that will, in turn, determine, whether customers are both US and EU citizens and to obtain a waiver from the EU customers who are US citizens to share the relevant data.

In short, what is required is that the best practice data governance program be tied into the compliance program. It is important to plan for the future and to design an effective data governance program to suit the organization needs with an operating model and organizational structures with clear, well-defined roles and responsibilities, flexible policies and processes in place, effective standards, metrics, and controls which enable organizations to respond and adapt to a revised version of existing regulations as well as new regulatory requirements.

Attaining compliance will require the joining of hands of business, IT, and data stakeholders from several business units and departments that have a stake in the compliance data. While by adopting a principle of good data governance, an organization should be in a position to comply with everything asked for by a regulator and compliance will, in essence, become a by-product of an existing and continuing process, it must be understood that data governance alone cannot help you be compliant.

With the vast amount of data that organizations store and process, there is a need for automation, scalability, and auditability through the deployment of appropriate tools and technologies as well as awareness and education program to help ensure that everyone across the organization is aware of the need and importance of compliance; they understand their specific roles, responsibilities, and data-related activities about compliance, the consequences of noncompliance, and their legal accountabilities towards data.

Data governance is an important element in ensuring compliance and compliance is one of the biggest drivers of data governance. Therefore, you cannot have one without the other. Data governance is not only about compliance and compliance is not only about data governance. There is more to it. For example, there are drivers of data governance like operational efficiency, data security, big data, analytics, and compliance also involves creating a culture of honesty and integrity, meeting high ethical, professional standards, policies, and processes not related to data.

It is important to achieve the right balance between compliance and performance (Lustig 2014). While compliance may be the motivator for data governance, the performance requirements also need to be kept in mind to ensure maximum/optimum results. A sole "checking the box" or 'tick box' approach to compliance without taking into account other drivers like performance, and efficiency may appear to be ok for a short term but will eventually bite you. In other words, you do not want to view data governance as a compliance activity only.

I will conclude this chapter with an interesting statement that Shannon Fuller, Director of Governance Advisory Services, Gray Matter Analytics made in his interview:

> Regulations are a big stick that you have to wield lightly as a governance practitioner.

References

AOGLLP https://www.aogllp.com/class-action-blog/taco-bell-agrees-to-25m-overtime-class-action-settlement#ixzz35dq9Y2EW Accessed June 7 2020

Baker S, Sjoberg P (2018) Intelligent data governance for Dummies®, Hitachi Vantara Special Edition. John Wiley & Sons, Inc.

BBC (2013) Barclays fined $3.75m after record-keeping failure, from BBC News. www.bbc.co.uk/news/business-25525621 Accessed 10 Jan 2018

BBC News (2018) Greenwich University fined £120,000 for data breach. BBC News-Technology. https://www.bbc.com/news/technology-44197118 Accessed 31 May 2018

BCBS Basel Committee on Banking Supervision (2013) Principles for effective risk data aggregation and risk reporting, © Bank for International Settlements 2013. https://www.bis.org/publ/bcbs239.pdf. ISBN 92-9197-913-9. (January, 2013)

BitSight Tech (2018) A risk manager's guide to the general data protection regulation (GDPR). www.bitsighttech.com

Blackmore N (2019) How to avoid pulling a Google and copping a $80 million fine for data misuse. Smart Company-Startup Advice. https://www.smartcompany.com.au/startupsmart/advice/google-80-million-fine-data-misuse-gdpr/ Accessed 20 May 2020

Brunelli M, Poor reference data management causing headaches in financial sector, TechTarget, E-guide, Bloomberg Vault. https://docs.media.bitpipe.com/io_10x/io_108208/item_640879/Bloomberg_sDataManagement_IO%23108208_E-Guide_021813.pdf Accessed 20 May 2020

Burgett A (2018) The cost of GDPR non-compliance: fines and penalties. KirkpatrickPrice Blog. https://kirkpatrickprice.com/white-papers/cost-gdpr-non-compliance-fines-penalties/ Accessed 10 Dec 2018

Carter J (2016) Beyond the Atlantic: data privacy laws around the world. Techradar.Com. https://bit.ly/2PuoCyR Accessed 26 Mar 2019

Chandler S (2019) Crypto is tightening up its anti-money laundering game, While Banks are still being fined for non-compliance. https://cointelegraph.com/news/crypto-is-tightening-up-its-anti-money-laundering-game-while-banks-are-still-being-fined-for-non-compliance Accessed 10 May 2019

Chatterjee D (2018) It's getting expensive not to be compliant. IBM RegTech Innovations. IBM Blogs. https://www.ibm.com/blogs/insights-on-business/banking/getting-expensive-not-compliant/ Accessed 10 May 2018

Cohen JK (2018) 3 major HIPAA fines so far in 2018, Becker's Health IT and CIO Report. https://www.beckershospitalreview.com/cybersecurity/3-major-hipaa-fines-thus-far-in-2018.html Accessed 10 Dec 2018

Collins A, Lightbourne J (2018) The U.S. Department of Health and Human Services Settles with Anthem for Record $16M Over Alleged HIPAA Violations, Cleary Enforcement Watch. https://www.clearyenforcementwatch.com/2018/10/u-s-department-health-human-services-settles-anthem-record-16m-alleged-hipaa-violations/ Accessed 10 May 2019

Cornerstone on Demand (2014) Manufacturing insights–5 workforce challenges and how top organizations solve them. https://www.cornerstoneondemand.com/sites/default/files/whitepaper/csod-wp-manufacturing-insights-072014.pdf Accessed 10 May 2019

DalleMule L, Davenport TH (2017) What's your data strategy? Havard Business Review. https://hbr.org/2017/05/whats-your-data-strategy Accessed 1 Dec 2018

Data Republic (2018) GDPR, CCPA and what they means for data governance-Data Republic

Difference Between (2019) Difference between Data Protection and Data Security. https://www.differencebetween.info/difference-between-data-protection-and-data-security Accessed 1 May 2020

erwin (2017) State of data governance report. https://go.erwin.com/2018-state-of-data-governance-report Accessed 28 Dec 2018

FDA et al (2018) U.S. Department of Health and Human Services Food and Drug Administration, Center for Drug Evaluation and Research (CDER), Center for Biologics Evaluation and Research (CBER), Center for Veterinary Medicine (CVM), December 2018. Data Integrity and Compliance With Drug CGMP: Questions and Answers Guidance for Industry, Pharmaceutical Quality/Manufacturing Standards (CGMP) https://www.fda.gov/media/119267/download Accessed 10 Jan 2019

Fearn N (2015) Carphone warehouse hit with 400k fine for 2015 data breach. IT Pro UK. https://www.itpro.co.uk/data-protection/30266/carphone-warehouse-hit-with-400k-fine-for-2015-data-breach Accessed 3 May 2018

Franco J-M (2018) BCBS 239, GDPR: Why must banks focus on data governance? Big Data and Digital Blog, Business and Decision. https://en.blog.businessdecision.com/bigdata-en/2018/02/bcbs-239-gdpr-must-banks-focus-data-governance/ Accessed 3 May 2018

Glenday J (2013) Sony fined £250k over 'serious' Data Protection Act breach. The Drum. https://www.thedrum.com/news/2013/01/24/sony-fined-250k-over-serious-data-protection-act-breach Accessed 10 Feb 2013

Gow B (2006) Case study: Data Governance & Compliance for Financial Services, IBM

Guess R (2011) What the Dodd-Frank Act may mean for DG in the financial sector. DataVersity. https://www.dataversity.net/what-the-dodd-frank-act-may-mean-for-dg-in-the-financial-sector/

Hall C (2016) Risk and regulatory reporting: centralisation is not the answer. Financial Markets Insights, The Realization Group. https://www.therealizationgroup.com/wp-content/uploads/2018/01/13-FMI-Risk-and-regulatory-reporting-160217.pdf Accessed 20 April 2018

[https://www.risk.net/risk-management/5781976/regulators-bristle-at-slow-progress-on-bcbs-239]

IAPP (2019) The international association of privacy professionals. Glossary of Privacy Terms. https://iapp.org/media/pdf/resource_center/IAPP_Privacy_Certification_Glossary_v2.0.0.2.pdf. Accessed 20 April 2019
ICO (2018) Bupa fined £175,000 for systemic data protection failures. https://ico.org.uk/about-the-ico/news-and-events/news-and-blogs/2018/09/bupa-fined-175-000-for-systemic-data-protection-failures/ Accessed 20 April 2019
ICO (2019) Information Commissioner's Office, The Guide to the General Data Protection Regulation (GDPR). https://ico.org.uk/media/for-organisations/guide-to-the-general-data-protection-regulation-gdpr-1-0.pdf Accessed 19 May 2019
Isidore C (2012) Standard chartered fined $327 million for violating sanctions. CNN Business. https://money.cnn.com/2012/12/10/investing/standard-chartered-sanctions-iran/index.html
i-Sight (2018) A practical guide to data privacy laws by Country. https://i-sight.com/resources/a-practical-guide-to-data-privacy-laws-by-country/
Jaclyn J (2008) Report: financial firms fined $26B for AML, sanctions, KYC non-compliance since 2008. Compliance Week. https://www.complianceweek.com/blogs/enforcement-action/financial-firms-fined-26b-for-aml-sanctions-kyc-non-compliance-since-2008 Accessed 1 Dec 2018
Keumars Afifi-Sabet NEWS (2018) Equifax hit with maximum £500,000 fine after massive security breach, ITPRO. https://www.itpro.co.uk/data-protection/31950/equifax-hit-with-maximum-500000-fine-after-massive-security-breach Accessed 1 Dec 2018
KPMG Advisory NV (2018) Enterprise data management-data retention. https://assets.kpmg/content/dam/kpmg/nl/pdf/2018/advisory/data-retention.pdf Accessed 31 Dec 2018
Lustig M (2014) Compliance at speed. O'Reilly Media, Inc
Maindarkar A, Thykatt R (2014) Seven steps to effective enterprise-wide compliance. Infosys White Paper
Martijn N, Hulstijn J, Bruijne M, Tan Y-H (2015) Determining the effects of data governance on the performance and compliance of enterprises in the logistics and retail sector, pp 454–466. https://doi.org/10.1007/978-3-319-25013-7_37
McDowall RD (2018) Data integrity and data governance: practical implementation in regulated laboratories. Royal Society of Chemistry
Merkulov P (2017) Data security & privacy: the risks of not playing by the rules. Dark Reading, Informa PLC. https://www.darkreading.com/risk/data-security-and-privacy-the-risks-of-not-playing-by-the-rules/a/d-id/1328951
Merkulov P (2018) The staggering costs of non-compliance, SC Magazine. https://www.scmagazine.com/home/opinions/blogs/executive-insight/the-staggering-costs-of-non-compliance/ Accessed 28 Dec 2018
Nilfisk (2016) Compliance for food manufacturers: the high cost of doing nothing. Industrial Vacuum Blog. https://news.nilfiskcfm.com/2016/10/compliance-for-food-manufacturers-the-high-cost-of-doing-nothing/ Accessed 02 May 2020
O'Hara M (2018) Risk and regulatory reporting: centralisation is not the answer. The Realization Group. https://www.therealizationgroup.com/portfolio/risk-and-regulatory-reporting-centralisation-is-not-the-answer/ Accessed 31 Dec 2018
OMB Memorandum M-07-16 (2007) Safeguarding against and responding to the breach of personally identifiable information
Ouellette P (2014) HHS deals out largest-ever $4.8M HIPAA violation settlement. Health IT Security. https://healthitsecurity.com/news/hhs-deals-out-largest-ever-4-8m-hipaa-violation-settlement
Pappu R (2018) Data governance–understanding its importance for AML compliance. ACA Telavance Blog. https://www.acacompliancegroup.com/blog/data-governance-%E2%80%93-understanding-its-importance-aml-compliance Accessed 28 Dec 2018
PCI SSC (2010) PCI security standards council, October, 2010. PCI DSS Quick Reference Guide
PharmaCompass (2017) 2016—A year of data integrity issues and pharma non-compliances. The PharmaCompass Newsletter. https://www.pharmacompass.com/radio-compass-blog/2016-a-year-of-data-integrity-issues-and-pharma-non-compliances Accessed 28 Dec 2017
Ponemon Institute LLC, Sponsored by Globalscape (2017) The true cost of compliance with data protection regulations. https://dynamic.globalscape.com/files/Whitepaper-The-True-Cost-of-Compliance-with-Data-Protection-Regulations.pdf Accessed 28 Dec 2018

Price E (2019) France Fines Google $57 Million for GDPR violations. https://fortune.com/2019/01/21/france-fines-google-57-million-for-gdpr-violations/ Accessed 1 Dec 2019

PricewaterhouseCoopers (2017) Anti-money laundering and counter-terrorism financing: from compliance to confidence. https://www.pwc.com.au/consulting/assets/anti-money-laundering-2017.pdf Accessed 1 Jan 2019

Raval T (2018) KYC and AML: What all banks need to know. Forbes Technology Council, Forbes. https://www.forbes.com/sites/forbestechcouncil/2018/10/11/kyc-and-aml-what-all-banks-need-to-know/#4bf65e1770fc

Reid D (2018) UK watchdog fines Facebook $644,000 for failing to protect its users' data. CNBC. https://www.cnbc.com/2018/10/25/uk-watchdog-fines-facebook-644000-for-failing-to-protect-user-data.html

Rodionova Z (2016) TalkTalk given record fine over data breach that led to data theft of nearly 157,000 customers. Independent News. https://www.independent.co.uk/news/business/news/talktalk-fine-data-breach-theft-customers-information-stolen-record-penalty-a7346316.html

Rouse M. Data retention policy. TechTarget. https://searchdatabackup.techtarget.com/definition/data-retention-policy

Spitzer J (2008) HIPAA through the years: 5 biggest fines since 2008. https://www.beckershospitalreview.com/cybersecurity/hipaa-through-the-years-5-biggest-fines-since-2008.html

TOC_HIT. The office of the national coordinator for health information technology guide to privacy and security of health information, Version 1.0 022112, https://www.healthit.gov/sites/default/files/pdf/privacy/onc_privacy_and_security_chapter4_v1_022112.pdf

Warman M (2013) Sony pays £250,000 PSN hack fine. The Telegraph. https://www.telegraph.co.uk/technology/sony/10180338/Sony-pays-250000-PSN-hack-fine.html

Westin AF (1967) Privacy and freedom. Atheneum, New York, p 7

Wiki-PCI-DSS. Payment card industry data security standard. https://en.wikipedia.org/wiki/Payment_Card_Industry_Data_Security_Standard

Yu E (2019) January 15, 2019, Firms fined $1M for SingHealth data security breach. ZDnet. https://www.zdnet.com/article/firms-fined-1m-for-singhealth-data-security-breach/

Chapter 6
Data Governance and Compliance: Concluding Thoughts and the Way Ahead

"Step by step, you make your way forward". —Gretchen Rubin

Abstract Governance has always been and will always be a "must have". Corporate governance and several sub-disciplines of corporate governance came into the picture with the evolution of corporations a few centuries back. This chapter summarizes the key milestones in the evolution of corporate governance, data and data governance, and discusses the way forward in brief.

6.1 Corporate Governance—Paving the Way

Governance has always been and will always be a "must have". Corporate governance and several sub-disciplines of corporate governance came into the picture with the evolution of corporations a few centuries back. Corporate governance and its sub-disciplines - finance governance, risk governance, HR governance, security governance, operational governance, data governance and IT governance have also evolved with evolution of technology and changing social and economic conditions. Table 6.1 shows the key milestones in the evolution of corporate governance, data, and data governance.

While finance governance, HR governance, risk governance, security governance and operational governance are relatively older subdisciplines of corporate governance, subdisciplines of corporate governance such as IT governance and data governance are relatively new. All these subdisciplines are tied together and feed into corporate governance and vice versa through exchange of information or data.

6.2 Data Governance Subdiscipline

The current age is the digital age with data being the king. Corporations are inundated with enormous amounts of data which keeps growing exponentially. With the

R. Mahanti, *Data Governance and Compliance*,
https://doi.org/10.1007/978-981-33-6877-4_6

Table 6.1 Key milestones in the evolution of corporate governance, data, and data governance

17th century	1st corporate governance dispute
19th century	Evolution of corporate governance acts
End of 19th century	Electronic data capture and storage
1930s	Great depression; Base for regulations and corporate governance
1960s	First record of the term "Corporate Governance"; Data seen as a by-product
1970s	Data governance practiced in some form
1980s	Corporate governance failures; Technology evolution to handle corporate governance failures
1990s	Corporate governance codes (as a result of corporate governance failures); Advent of Internet; Data security challenges
2000s	Corporate governance scandals; DG gaining attention—Compliance and risk act as main drivers of DG
2000 onward	Evolution of treatment of data as an asset; More consensus around DG; Scope of DG expanding to meet data complexity, regulations, security requirements, etc.
2001–2002	SOX
2004	Basel-II
2008	Global Financial Crisis (GFC)—led to increased regulations
2018	GDPR

vast amounts of diverse data that corporations amass, with internet creating new data security challenges and need to manage data efficiently in today's highly competitive and regulatory corporate environment, where stakes are extremely high, data governance is a "must have" to increase operational efficiency, manage risks, stay compliant and maximize value. While corporate governance is a part of the corporate strategy, data governance is a part of the data strategy.

6.3 Regulations, Compliance, and Data Governance

The evolution of corporate governance and its subdisciplines have been influenced by the advent and evolution of regulations. These regulations have evolved in the face of corporate failures and scandals. Corporate accounting failures have led to financial reporting regulations which requires reporting finance data at a more granular level and asks for traceability of data to authoratitive source. Large amounts personal and sensitive data collected by organizations and there leakage and misuse (for example identity theft and fraud) have led to the creation of data privacy, data security and data protection laws. Data has evolved from being a by-product to being an asset in organizations that realize the impact and value of data. Data being an asset will be discussed in detail in the second book in the series–Data Governance and Data Management.

Compliance is one of the biggest drivers of data governance which we discussed in detail in this book. It is important to understand the regulations, the underlying data that the regulations have a direct and indirect impact on and the various data aspects and activities that need to be undertaken to stay compliant.

Regulations are lengthy complex documents and in order to correctly and completely map regulatory requirements to the correct data elements, organization need to have a cross functional team consisting of stakeholders and subject matter experts from several departments and business units–Compliance, Risk, Legal, impacted business units and IT (teams which handle the data impacted by the regulations). With the huge amounts of data that organizations collect and store in different data repositories across the organization, locating the compliance related data is difficult and time taking.

The entire exercise to achieve compliance is lengthy, complex, as well as extremely expensive. Nonetheless it needs to be undertaken; this is because, while the cost of compliance is high, the cost of non-compliance is way higher. Data governance plays a key role in achieving compliance.

6.4 The Way Forward

While compliance is one of the major data governance drivers, there are other data governance drivers too, which push for the implementation of data governance. These are reputation management, improving customer satisfaction, better decision making, improving data quality, data security, data privacy, data analytics, big data, improving operational efficiency, revenue growth, mergers and acquisitions (M&A), partnering, and outsourcing. Data governance has several benefits too such as increased operational efficiency, improved data understanding, improved data lineage, better decision making, and greater data quality, increased confidence in the data, improved risk mitigation, and cost savings. All these drivers and benefits will be discussed in the second book in the series–Data Governance and Data Management.

Data governance ties together several data management disciplines and data initiatives–data architecture management, data modeling and design, data quality management (DQM), data security management, data warehousing and business intelligence (BI) management, data integration and interoperability (DII), document and content management, metadata management, reference data management, master data management (MDM), data storage and operations, and big data, data lake and data analytics. These will be discussed in the second book in the series–Data Governance and Data Management.

Chapter 7
Data Governance Interviews

Abstract As discussed before, interviews were conducted to elicit the experience and unique perspectives of highly experienced professionals—thought leaders, researchers, and consultants for the DG book trilogy entitled *Data Governance: The Way Forward*. These people have shared their thoughts on data and data governance. The questions asked and their respective responses are presented in this chapter.

As discussed before, interviews were conducted to elicit the experience and unique perspectives of people—thought leaders, researchers, and consultants. These people have shared their thoughts on data and data governance. The professionals who gave interviews for the DG book series include:

1. Andres Perez, Information Management Consultant and President, IRM Consulting, Ltd. Co.
2. Christopher Butler, CDO HSBC UK, HSBC
3. George Firican, Director, Data Governance & Business Intelligence @UBC DAE | Founder of www.lightsondata.com
4. Jill Dyché, Principal, Jill Dyché, LLC.
5. Dr. John R. Talburt, Acxiom Chair of Information Quality at the University of Arkansas at Little Rock, and Lead Consultant for Data Governance and Data Integration with Noetic Partners Inc.
6. John A. Zachman, John A. Zachman, Author of "The Framework for Enterprise Architecture" (The "Zachman Framework"), Zachman International
7. Laura Sebastian-Coleman, Data Quality Lead, Finance Data Governance Center of Excellence, CVS Health
8. Phil Watt, Director, Elait Australia
9. Shannon Fuller, Director of Governance Advisory Services, Gray Matter Analytics
10. Dr. Stan Rifkin, Director of R&D, Master Systems Inc.
11. Tony K. Epler, Chief Data Strategist, PricewaterhouseCoopers

R. Mahanti, *Data Governance and Compliance*,
https://doi.org/10.1007/978-981-33-6877-4_7

7.1 Andres Perez, Information Management Consultant and President, IRM Consulting, Ltd. Co.

Andres Perez is an Information Management Consultant and President of IRM Consulting, Ltd. Co., based in San Antonio, Texas. Mr. Pérez specializes in information resource management, information architecture, and information quality management.

1. What is the biggest business driver of data governance?
   Organizations are using more data and increasingly sharing data across lines of businesses that traditionally were kept separate and, in doing so, realizing that the data definition and content are not meeting their business needs. Also, organizations are beginning to understand that data is a valuable resource that must be managed well to get better outcomes.
2. What is the biggest challenge you have ever faced in a data governance implementation that you have worked on?
   There are two "bumps" on the road that typically cause data governance derailments. One is the resistance by IT groups to open up and share the necessary information about their systems, practices, and data so that we can understand how the data is acquired, stored, and used by critical applications.
   The other is, once DG is accepted, that organizations continue working with a "business as usual" mindset, and the responsible individuals continue behaving as they did before, failing to understand that they are no longer working for their stovepipe team but for the enterprise across the information value chain.
3. What, in your opinion, are three most important factors that are necessary for success of a data governance program?

   a. Leadership: there must be a business (not IT) leader that helps other business leaders that information has values when applied and has cost when defined (planned), acquired, stored, and disposed of. This individual must understand the critical pain points across business areas and must be able to relate it to others.
   b. Organizational Structure: most organizations fail to erect the proper organizational capabilities to facilitate managing information across its value chain; they tend to focus on technologies and are swayed by vendors into believing that the "silver bullet" technology they provide is all they need.
   c. Training: data governance is difficult. It goes against the grain of the typical organization. Organizations are managed top-down using a command and control (hierarchical) method while the information flows across all of these stovepipes. Data governance must be practiced by all members of the organization, and the DG leadership must provide proper training to them so that they can work across organizational borders effectively.

4. Why do many data stewards fail?
   The job of the data steward is complex and requires proper leadership (see 3.a above), organizational capability (see 3.b), and training (see 3.c). The

complexity comes from the fact that a steward must learn to work across the organization with a keen focus on collaboration and compromise (negotiation).

5. How does data governance contribute to a data-driven culture?
   It does when it is effective. It should provide a beacon for data value realization. The value is not distributed evenly across business areas, so it is important for those who gain the most to share the wealth with those who do not by providing feedback on business value and by treating each other as partners (a common "talking point" but not a common practice in business today).
6. Any other interesting thoughts/advice that you would like to share with the audience of the book with respect to data governance?
   Data governance is not new; data stewardship is not new. Some organizations began these activities since the late 1970s. What is new is their widespread practice leading to better understanding, shared knowledge, and better tools (collaboration tools and catalogs).

We are beginning to break away from (using Auguste Comte's evolutionary model) the ***theological stage***, in which data was thought as mythically conceived and explained in dogmatic ways. Now, we are in the ***metaphysical stage*** in which data is perceived as an obscure force resulting from the "natural ways in which organizations operate" explained in obscure ways. I hope that what we learn next will lead us into the ***positive stage*** in which all abstract and obscure forces are discarded, and the data phenomena are explained by real and objective (value based) means.

7.2 Christopher Butler, CDO HSBC UK, HSBC

Christopher Butler is the Chief Data Officer of HSBC, UK Bank plc. He has demonstrated thought leadership across all aspects of data management and the effective use of data to achieve business outcomes. He has successfully led both transformational and operational teams at a global, regional, and country level.

1. In your opinion, why is there no standard definition around data governance?
   Governance within an organization has been around a long time, and the concepts are well understood. Data proliferation, use, ease of sharing, data privacy, increasing data regulation, and businesses' understanding of how data can be used to drive growth is expanding exponentially. It is the successful merging of these areas within an existing organization's structures or effectively changing the organization's structures to take advantage of data that provides the challenge. I believe that the lack of success of many organizations to do this successfully brings about the opinion that no standard definition around data governance exists, where in fact, clear standards around governance and data management do exist today.

2. Why do data governance programs fail?
The focus is on data governance as a standalone objective without considering the broader business objectives and how strong data governance can support these. Even with *top down* support, any data governance program must be in conjunction with other business & IT initiatives.
The program must be fully understood, funded, and be able to demonstrate long-term sustainable benefits.
3. In your opinion, how important is the Chief Data Officer's role in the success of data governance?
The Chief Data Officer is critical to aligning business and data areas together. He must fully understand and be part of the business structure, provide the necessary data governance as well as alignment with IT tools and technology to move forward.
The CDO is able to ensure the right level of 'C' Suite support and be in a position to ensure that consistency and standards apply to the entire organization.
4. How has technology facilitated the implementation of data governance?
As with all technology—these are tools that are used to measure, report and manage data as part of the governance process. However, what you measure and record is more important than just having a tool in place. The information, dashboards, and processes must all fit within a holistic environment and be adapted to demonstrate value.
5. How have regulations impacted data and influenced data governance?
Overall, increasing regulation has had (and will continue to have) a massive impact on data management (Basel II, BCBS 239, Liquidity, MiFID). These have made organizations much more aware of the data and information they have available. However, it should be noted that there are many other factors also driving the increase in data governance - from cybersecurity, data privacy, digitalization, and customers expecting more.
6. Any specific advice that you would like to share with the audience of the book with respect to data governance?
Data governance should be seen as an extension to existing practices and not a new concept. Start small with the tools that you have and align with existing business practices. The key aim is education around data and data matters. The business world is changing. A business that has embedded robust data management practices into its processes to ensure quality data is in a better position for long-term survival. This can be through making better business decisions, operating efficiencies, identifying new products/customers or servicing the customers you have today.
The one thing for certain is that data will become more and more important as a differentiator for businesses going forward. Data governance is at the heart of whether an organization will be in a position to benefit from its data.

7.3 George Firican, Director, Data Governance & Business Intelligence @UBC DAE | Founder of www.lightsondata.com

George Firican is a Data Governance and Data Management practitioner currently working with The University of British Columbia, which is ranked among the top 40 public universities in the world. His innovative approach to data management received international recognition through award-winning program and project implementations in data governance, data quality, business intelligence, and data analytics. As a passionate advocate for the importance of data, he is a frequent conference speaker and YouTuber and has been ranked among the Top 35 Global Thought Leaders and Influencers on Digital Disruption and among the Top 15 on Innovation and Big Data.

He advises organizations on how to treat data as an asset, and shares practical takeaways on social media, different industry sites and publications, and is the founder of https://www.LightsOnData.com.

1. How long have you been implementing data governance programs and how has data governance evolved in your view?
   I have been implementing data governance programs since early 2012, but recognized their importance since I wrote my first database application.
   From my experience, data governance used to be considered solely an IT function, usually being siloed from the rest of the organization. Then, as the usefulness of data became more tangible and organizations realized the power of being data-driven, data governance also became a more collaborative function, between the business and IT.
   Fast-forward to today, data governance is no longer an option, but it's essential to be implemented and adopted not just organization-wide, but world-wide. Noting that "data drives everything", Japan's Prime Minister declared earlier in 2019 that he "would like Osaka G-20 to be long remembered as the summit that started worldwide data governance." That's quite the evolution for data governance.
2. What is the role of strategy and business case in paving the way to achieve success in data governance?
   The business case is needed in order to secure sponsorship for a data governance program. The overall strategy needs to address the business needs, deliverables, objectives, framework, and best practices as well as the change management accompanying it all. Both the strategy and business case should help identify the risks, gaps, assumptions, constraints, and benefits, tie them to the goals of the organization to determine why and how data governance can support them.
3. How do you associate the data governance maturity levels with business benefits?
   There is a direct correlation between the maturity of the data governance program and the business benefits. At early maturity stages, organizations reap few, if any, benefits from their data. As data governance emerges, ROIs become

more apparent, until the highest maturity state of the program when the organizations only rely on data-based decisions and treat this asset as a means to gain a competitive advantage.

4. What factors should be taken into consideration while choosing a data governance model for an organization?
   Assuming you are referring to data governance operating models: the size of the organization (ex: small organizations typically benefit from a centralized model), the organizational structure and that of its operations (ex: a decentralized model can work well for an organization which has dispersed its operations to several remote locations) the main drivers, and complexity of its business and technical environment (ex: that with a high level of complexity might adopt a federated model).
5. What role does governance organization structures play in the successful implementation of data governance in an organization?
   Data governance has its share of politics and change management struggles. The program structure needs to be shaped in such a way to best address these struggles and mitigate the risks associated with them.
6. Any other interesting thoughts/advice that you like to share with the audience of the book with respect to data governance?
   Data is an important asset and always needs to be treated and managed as such. Data governance programs are essential in managing this asset. After all, the engine for economic growth is no longer fuelled by gasoline but more and more by data.

7.4 Jill Dyché, Principal, Jill Dyché, LLC.

Jill Dyché has been thinking, writing and speaking about data and analytics for over two decades. In her career as a management consultant and executive advisor, she's helped companies deliver strategic change through data and analytics. And as a popular author, she has both educated and entertained her many readers.

Jill was a co-founder of Baseline Consulting Group, an analytics and data strategy firm. At Baseline, she combined the roles of best practice expert, industry gadfly, executive coach, and business owner.

In 2011, Baseline was acquired by the pioneer analytics vendor SAS, where Jill served as Vice President of Best Practices, leading a team of senior specialists advising customers in success factors for analytics, data management, and AI. She is the author of four books on the business value of technology, including the award-winner, The New IT (McGraw-Hill).

Jill is a popular blogger and keynote speaker. She is also the Executive Director of Outta the Cage, a non-profit that rescues and rehabilitates at-risk shelter dogs. Reach her at jill@jilldyche.com.

1. How long have you implemented data governance and how has data governance evolved in your view?

I started working with clients in data governance back in the mid-2000s. Back then, every client defined data governance differently. Some saw it as a central board of decision-makers. Others considered it to be synonymous with data quality. Now there's much more consensus on what data governance is, and how it should be deployed.

2. What is the role of strategy in paving the way to achieve success in data governance?
   The most sophisticated companies use strategy to determine their data deployment priorities. In other words, there's a lot of data and a lot of potential data-enablement projects, so where should we start? The short answer is to consider your company's most important strategic initiatives and figure out the data that will be required to execute them. Start there.
   I spend most of my time doing strategy consulting these days, and it's fascinating how data is still a HUGE part of these conversations! From a corporate strategy perspective, we use a strategy mapping technique to map corporate objectives to the data necessary to achieve them. This approach essentially puts data—and the need for data governance—on the radar of executives who might believe data to be an afterthought. Executives, many of whom are measured on strategy execution, will typically support data governance as soon as they understand its role in strategic enablement.
   The extent to which companies can link their corporate strategies to data is the extent to which they'll get support and funding for data governance!
3. What is the role of data governance framework and data governance operating models in the successful implementation of data governance?
   Critical. Data governance frameworks and delivery models help to not only define data governance, they establish its boundaries and success measures. If delivery models are absent, data professionals can end up "going rogue" and fixing things that aren't broken. Frameworks keep people on track and ensure that there's a delivery pipeline.
4. What factors should be taken into consideration while choosing a data governance model for an organization?
   Culture, culture, culture! Understand how successful projects are delivered at your company. Is it largely top-down, meaning that executive sponsorship is required and there should be formalized budgets for each initiative? Or is success typically more bottom-up, where a group of data management experts deploys valuable new data using agile techniques, then rinse and repeat?
   Understanding how the culture defines success will inform not only how you structure data governance but who you hire.
5. What skill sets, expertise, and knowledge are important in the implementation of data governance?
   This one goes back to priorities. Many companies won't agree to formalize data governance until there's an authoritative platform for targeted data. Others require proofs of concept—often for data correction or data integration efforts—before being willing to launch a data governance team.
   At a minimum, I'd say the necessary skills fall into three buckets.

1: Business analysts. Every company needs data-conversant analysts who can define data requirements and tie them back to business requirements and strategy.
2: Data stewardship. Tactical data stewards can ensure that data quality, metadata, data modeling, and data deployment are formal processes using best-of-breed toolsets.
3: Platform specialists. Whether you're relying on a legacy data warehouse or a newfangled big data in the cloud platform, there should be an understanding of data's sources and destinations. Designating specific platforms for specific purposes leads to trustworthy data.

6. Any other interesting thoughts/advice that you like to share with the audience of the book with respect to data governance?
   Data governance is changing. It's no longer a separate effort but rather it's "baked in" to every strategic project. The good news is that data has a seat at the strategy table now. It should have been invited sooner!

7.5 Dr. John R. Talburt, Acxiom Chair of Information Quality at the University of Arkansas at Little Rock, and Lead Consultant for Data Governance and Data Integration with Noetic Partners Inc.

John R. Talburt, PhD, IQCP, CDMP, is the Acxiom Chair of Information Quality and Director of the Information Quality Graduate Program at the University of Arkansas at Little Rock. Previously, he led the Data Research and Development group for Acxiom Corporation where he implemented their Total Data Quality Management program. He is currently serving as a member of the U.S. Technical Advisory Group (TAG) to the International Organization for Standardization (ISO) in the area of data quality. He is an inventor for several patents related to customer data integration and the author of numerous research papers and articles on information quality and entity resolution. His books include *Entity Information Life Cycle for Big Data: Master Data Management and Information Integration* (Morgan Kaufmann 2015), *Entity Resolution and Information Quality* (Morgan Kaufmann, 2011), *Data Engineering: Mining, Information and Intelligence* (Springer, 2010), and *Information Quality and Governance for Business Intelligence* (IGI Global, 2014).

1. Why is data not treated as an enterprise asset?
   I think this has changed significantly over the past couple of years. I agree traditionally data was thought of as a commodity and all of the focus was on programming and systems. This is probably a natural evolution. It has taken a long time to tame the program development cycle and produce re-usable, open source code. With the advent of distributed computing and cloud platforms, we are seeing a reversal of this attitude. Programs and systems now are becoming

more of a commodity, and organizations are realizing their competitive edge lies in the use of their own data resources.

Now organizations are actively trying to manage data as an asset and a resource, thus, we are now seeing a rising interest in data governance as an important part of the solution. You can see a growing number of organizations beginning to embrace data governance and the concept of data as an asset.

One problem I see is that they want to jump to the end of the data curation process and immediately start using clever and sexy data analytics in the hopes of achieving break-through insights. They are attracted to all of the buzz about using AI and machine learning, while skipping over the hard part of data governance, data integration, and data quality.

In my opinion, DG success or failure really depends on the involvement of the business side of the house. Organizations viewing DG as an IT program are doomed to failure in both DG and analytics efforts. The real opportunity with DG adoption is not only derived from more effective data management, but just as important, getting business and IT engaged with each other. Even though data resides in IT systems, it is the business who uses data to create value. In a good DG program, the business defines "what" they want to happen with data and engage IT to help define "how" it will happen. I have seen an IT group breathe a sigh of relief when the business actually assumed this role. Many IT groups are overburdened with accountability for data they are not really using. Business should be accountable for data, and IT for networks and systems, and DG is where they find common ground.

2. Why and how does big data make data governance harder?

   Big data is a double whammy. At the same time, organizations want to actively manage data as an asset through data governance but they also need to move from Mode 1 (traditional RDBMS client/server) to Mode 2 (HDFS platform) computing to take advantage of advanced analytics and high-volume processing capabilities.

   My observation is that these two things are tightly coupled. Whereas DG was important to efficient data management in Mode 1, it is absolutely essential in Mode 2 computing. The basic tenets of DG are "know what data you have, what's in it, and where it's located". Big data not only brings large volumes of records per dataset, it brings large volumes of different datasets.

   Keeping track of these datasets and understanding their content is difficult using the Mode 1 approach which starts with a pre-defined data model. In Mode 2, organizations need to ingest new data quickly without using an ETL process to map data into a model. At the same time, they still need to track what's in the data and know where it's located, but now for more datasets and more quickly. Mode 2 computing puts even more pressure on DG. Many have observed, Mode 2 computing using the raw data ingestion ("data lake") approach without good DG results in a "data swamp".

3. Can you throw light on some important metrics that should be gathered to measure the success of a data governance program?

In my experience, I see DG becoming more mature. The first adopters usually had to make 2, or even 3, attempts before seeing any success. However, like any new paradigm it takes time to iron out the wrinkles. As more and more organizations implement DG and share their "lessons learned," DG practice is improving and success rates are increasing. We now have good models to follow. It is still a very difficult undertaking but I see things beginning to get better with fewer failures and restarts.

As with any program, DG can't be effectively managed without measurement. I encourage organizations to develop three types of DG metrics—progress, verification, and validation metrics. Progress metrics track the development of the DG program, e.g., number of datasets/element catalogued, number of standards developed versus target, and how many data stewards have been trained. Verification metrics track the degree to which the program is actually working as designed. Are employees and processes actually conforming to policies and standards? Verification metrics are usually measured by policy and standards reviews of internal projects and initiatives. Verification metrics are very much DG quality measurements in the sense that they measure the degree to which data management processes conform to DG policy and standards, just as data quality is the degree to which data conform to data quality requirements.

Finally, validation metrics are some of the most important measurement, but unfortunately, the measurements are often left out. Validation metrics measure the degree to which the business case for DG is being achieved. Every DG program should start with a good business case projecting the value of the program. It is very important that these business objectives be tracked and measured. For example, if one of the value propositions was that the DG program would reduce regulatory fines by 50%, then a validation metric would answer the questions, is that objective being achieved? In my view, a good DG dashboard should include all three types of metrics.

4. How has technology facilitated the implementation of data governance?

In general, I subscribe to the axiom "A new system is not the answer." Most obstacles in the adoption of new computing paradigms are "people problems," not technology issues. Viewing and managing information as an enterprise asset will require the same kind of cultural change that was needed for moving from procedural to object-oriented program, data warehousing, and now the shift to HDFS-based applications and cloud computing. As noted earlier, one of the great benefits of DG when implemented properly is to strengthen the collaboration between business and IT to bring them into better alignment around business goals and objectives.

At the same time, it is well-known in the DG practitioner community that the most commonly used DG tool is Microsoft Excel. While this approach might work to get DG started, or even suffice for a small organization, it does not have enough capability to sustain a robust DG program in larger enterprises. One of the greatest problems we face in DG now is the lack of metadata automation. Even though we now understand how important it is to curate and track all data across the organization, actually doing it is still largely a manual operation.

While many software companies are beginning to address these problems and are producing DG specific software tools, we still have a long way to go. Large enterprises need an army of data stewards to describe, classify, and manage their data assets. Because so much of traditional computing is only focused on the process and not on the data, tracking where data inputs come from, how the data are transformed, and where the outputs go has largely been left to be recorded and managed by people.

Here is where I see another tight coupling, the coupling between metadata and automation. Most systems don't create and don't use enough metadata. This is the legacy of expensive processing and storage in the past. In the early days, there was a great deal of focus on making data as compact as possible. The result was the decoupling of metadata from the data. Whereas the data was in a machine-readable file, the metadata describing the file was typically on a sheet of paper if it existed at all. Personally, I don't believe we will see great strides in DG automation until software systems start generating, storing, and using metadata with the data. For now, we seem content to have a few hundred bytes of layout information to describe a file of 10 million bytes.

This paradigm really needs to change. We need to think in terms of thousands of bytes of metadata for each byte of data. Just think of the business and technical descriptions and requirements, data movement and provenance, data classification, data access control, data quality, and the many other metadata relevant to just one item of operational data.

Even with more metadata stored with the data, I believe the entire paradigm of data processing needs to change to make DG effective. We need to think more like network security. If you think about it, every enterprise has a "network firewall" to manage and control every in-bound and out-bound network connection. To achieve automation in DG, we need to follow a similar pattern and build a "data firewall" around enterprise data. There needs to be a layer between data and all data users including developers. This layer is sometimes called a "positive data control" or PDC layer.

A PDC layer would enable and manage connections between user processes and data. Users, including developers, would no longer be able to directly connect programs to data and directly store the results. Instead, they would connect to the data through the PDC layer. All programs and processes would have to go through and be controlled by the PDC layer to read and write data. In this way, the layer would always be aware of everything that happens to every data element. Where it was read from, what process used it, and where it was written? The key component of the PDC layer is a smart data catalog to store all of this metadata.

Several software companies are already moving in this direction to develop more and more sophisticated data catalogs but so far, PDC has only been embraced for "end user analytics" systems developed to allow certain groups of business users to setup and run analytic processes without being trained as programmers. PDC would be a universal adoption of this model, not just for business users but for all data users, and yes, this would include system developers as well as business users.

5. In your opinion, how important is the chief data officer role in the success of data governance, and what are his/her responsibilities?
The CDO role can be very successful when implemented correctly. Academic research has already established this fact. Unfortunately, many organizations still make IT accountable for data. As noted in my earlier comments, the CDO and a DG program will not be successful without business participation. The most effective CDOs report to a business leader such as the CEO, CMO, CRO. When a CDO reports to the CIO, it is a sign that the organization does not understand the need for business to be accountable for data, and the CDO role is much less effective. As long as organizations see IT as accountable for data and DG, they will at best have limited success.
Another thing I often see in my work is the view of the CDO as mainly responsible for DG and DQ programs. Even though the CDO should definitely oversee DG and DQ, the real value a CDO can bring to an organization is leadership in developing a data strategy and vision. The CDO needs to be an executive-level leader outlining how the organization will best exploit its data assets and how it should partner with others to combine data for new products and services. Viewing the CDO as purely operational role for DG and DQ limits the role, and can cause organizations to miss out on a large part of the opportunity to participate in the data revolution.
6. Any other interesting thoughts/advice that you would like to share with the audience of the book with respect to data governance?
Data governance is clearly growing in adoption and maturing in practice. Having been around for many years, I have seen many shifts in paradigms around computing. For example, one of the great benefits of the data warehousing movement was not just to create business intelligence (BI) but also bring focus on the problems of information and data quality.
I see the same thing happening with DG. One of the greatest benefits of DG is not just more effective data management, it is also helping to bring business and IT together in a more synergistic and comfortable partnership. The truth is, I often find the relationship between business and IT to be very strained. They are often at odds with each other, each blaming the other over a lack of responsiveness and issues of control. The implementation of the DG program is a great opportunity to bring business and IT into a more harmonious and productive partnership.

7.6 John A. Zachman, Author of "The Framework for Enterprise Architecture" (The "Zachman Framework"), Zachman International

John A. Zachman is an American business and IT consultant, early pioneer of enterprise architecture, and Chief Executive Officer of Zachman International (Zachman.com). He is a thought leader and the originator of the "Framework for Enterprise Architecture" (The Zachman Framework™) which has received broad acceptance around the world as an integrative framework, an ontology for descriptive representations for Enterprises. Mr. Zachman serves on the Executive Council for Information Management and Technology (ECIMT) of the United States Government Accountability Office (GAO) and on the Advisory Board of the Data Administration Management Association International (DAMA-I) which awarded him the 2002 Lifetime Achievement Award.

1. What is the role of strategy and business case in paving the way to achieve success in data governance?
   There is an old engineering adage, "Form Follows Function." Today, this might be restated as "Structure Follows Strategy." Structure (Form) takes time and costs money. If you expend the time and money to build it before you define what it is for or what it is to do, you have delivered a classic solution in search of a problem … which in all likelihood is a huge waste of resources.
2. What is the role of data governance framework in the successful implementation of data governance?
   The word "governance" has recently become a popular replacement for the term "management." Maybe "governance" implies something more "authoritarian" or maybe more "structural" or something but the definition of "Management" historically included: 1. Plan, 2. Organize, 3. Staff, 4. Direct, and 5. Control. (sometimes supplemented by 6. Evaluate, (separated from Direct) and 7. Motivation (separated from Control). I am not familiar with the "data governance framework" but if it did not include at least these 5 components of "Management" I would look for an explanation or some rationalization for the omission.
3. How do you associate the data governance maturity levels with business benefits?
   I presume that the data governance maturity levels are those defined by the Capability Maturity Model (CMM) pattern including 1. Initial 2. Repeatable 3. Defined 4. Capable and 5. Efficient. In this case, I would associate the business benefits of the data governance maturity levels precisely the same as the business benefits of the Capability Maturity Model pattern that is extensively documented in Technology literature beginning with "Managing the Software Process" by Watts Humphrey.
4. Which industry sectors in your opinion have been most lax in implementing data governance and why?

I do not have specific experience to answer this question. However, there are likely patterns that would be helpful in identifying lax industries and rationalizing their laxity state. I will suggest a metaphor for illustration. Back in the '70's timeframe, the oil industry was thriving such that they could "spill more oil than they sold and still make a profit." Any industry where they are making so much money or getting so many tax dollars (which would include Public Sector Departments) that they did not need efficiency, has no incentive to invest in infrastructure, that is, anything that is "indirect" in an accounting sense or not required in a regulatory sense.

5. Which industry sectors have embraced data governance the most and what are their drivers for implementing data governance?
   The answer to this question is the converse of the answer to question 4. Any industry (or Public Sector Department) that is struggling or under heavy public scrutiny is a candidate for investing in infrastructure, particularly infrastructure that enhances data quality. Maybe, health care is a reasonable current example.
6. Any other interesting thoughts/advice that you would like to share with the audience of the book with respect to data governance?
   The big problem with data is semantic coherence, typically referred to as "data quality." If the data does not accurately reflect reality or is being conceived or understood inconsistently, then there WILL BE errors, defects, mis-understandings, conflicts, war, and/or untold costs of entropy of reconciliation. I just acquired a book "The Real Work of Data Science" which may hold valuable advice addressing this issue. One of the authors of this book is Thomas C. Redman, a friend who has given this issue of data quality more thought than any other human being I know. I clearly do not know every human being but at 85 years old, I know a LOT of them. None to my knowledge holds a candle to Tom Redman when it comes to data quality. He not only is thoughtful, logical, rational, and articulate … he is practical. I am anxious to read this new book. It does not specifically address the subject of "data governance," but it clearly addresses its raison d'etre.

7.7 Laura Sebastian-Coleman, Data Quality Lead, Finance Data Governance Center of Excellence, CVS Health

Laura Sebastian-Coleman is the Data Quality Lead of Finance Data Governance Center of Excellence at CVS Health in the US. She is an experienced data quality practitioner, author, and speaker with in-depth knowledge of health care data and data management practices. The author of two books–Measuring Data Quality for Ongoing Improvement and Navigating the Labyrinth: An Executive Guide to Data Management—Laura was production editor of DAMA's Data Management Body of Knowledge, second edition (DMBOK2).

1. While data is an enterprise asset, it is not treated as such. What are the reasons behind this?

Many factors limit the ability of organizations to treat data as an asset. First, in many organizations, data is created as a byproduct of transactional processes, rather than something that has value for other uses (analysis, planning, etc.). While lots of organizations now claim to be or strive to be "data-driven", most still do not think about data production in terms of the quality of the data. Secondly, people have trouble abstracting a monetary value from data. If asked the question "How much is our data worth?" most CEOs probably would not be able to produce a consistent answer. Finally, there are still misconceptions about data itself. Even very smart people stumble around when trying to think about data outside of technology. This means that conversations that should be about data end up being about tools and IT processes, rather than data.

2. What are the biggest challenges that you face when implementing data governance?
   Since many organizations have not governed their data in the past, at least at an enterprise level, data governance is perceived as a "new" thing that is making new demands on people's time. This means it requires that people change their behavior. So, data governance always requires some level of culture change.
   Culture change will look different in different organizations because the starting point will be different. The behaviors that need to change will differ, but the behaviors that people will be encouraged to adopt will be similar (e.g., simply paying attention to data and asking questions if it does not look right is a behavior to be cultivated in any organization that wants to govern its data better).
   The best definition I know of "culture" is "the ways people work together." Culture is about who is responsible and accountable for different activities, as well as about whom you can depend on to get things done. Asking people to behave differently around data can make new demands on all these areas.
   One of the biggest challenges I have faced (in three organizations) is describing to people, in the language of the organization, the behaviors they should adopt. This is followed by actually getting them to change those behaviors. Unfortunately, there are obstacles to culture change that individuals cannot always overcome. For example, IT funding models can be direct obstacles to data quality if those models focus solely on meeting project timeframes and do not account for the quality of the data produced by projects.
3. How do you manage and resolve potential organizational conflicts from overlapping functions when dealing with enterprise data?
   In my past experience, organizational conflict over data has been a big challenge. This has not been about who owns the data or who uses the data. It has been about who has to pay if the data is not right.
   Conflict itself can be reduced through good facilitation; especially when this is focused on ensuring that all parties have as objective a view of a problem's costs and benefits. But what actually gets done depends largely on funding.
4. How has HIPAA influenced your data governance program, and how has data governance helped meeting HIPAA requirements?
   HIPAA has been in place since before there was a strong concept of "data governance". It is a good example of a challenge that required organizations to

figure out how to organize decisions and solutions around data at an enterprise level.
Right now, within Health Care in the US, the focus is less on HIPAA per se, and more on ensuring that personal data is protected for other reasons, such as malicious actors. There is also simply a lot of data in a lot of different places within any health care organization, and as we move to different data storage platforms (e.g., Hadoop and other big data platforms), that data may be at risk if we do not inventory it and ensure that it is protected.

5. Any recommendations on metrics that should be used to measure the success of a data governance program?
In my current role, we are in the process of defining our KPIs. These are aligned directly with our strategic plan and are designed to show progress on foundational pieces of that plan. So, we have focused on the rate at which we are getting critical data defined, the coverage of our data quality metrics, and the processes for providing data consumer support. As we operationalize our governance processes, we will produce operational KPIs.
One reason why we are focusing on producing metrics is that, without them, other people do not have a clear picture of what the data governance team is actually accomplishing. With them, we can show that we are making headway on our initiatives. People get more interested when they feel like there is something concrete to learn about.
6. Any other interesting thoughts/advice that you would like to share with the audience of the book with respect to data governance?
I think we are heading into a new era of data governance. The existing assumptions and models for data governance have not produced the results that people have expected. There is still a limited understanding of the risks associated with data in organizations. Until we face these more squarely, I do not think that we can progress in the ways that we want to. Even the idea that data is an asset still has not resulted in improvements in how we manage data. People are more focused on technology than on data. So… be aware as you engage in this work, that it takes a lot of persistence and a willingness to try new things.

7.8 Phil Watt, Director, Elait Australia

Phil Watt is a Director for Elait Australia, specializing in data engineering and data governance solution services. He has more than 25 years' experience in large scale enterprise analytics and data engineering programs, and his interest is in the journey to scale business programs from small, proof of concept initiatives through to operational company-wide solutions with a high strategic impact.

1. How long have you been implementing data governance programs and how has data governance evolved in your view?
My first data governance program was in BT (British Telecom around 2002). Over the next 10–15 years, little changed in the literature and standard

approaches. Large scale top-down approaches characterized with data stewards and significant manual effort were advocated and adopted in many large organizations and almost always with disappointing results.

Over this time I had pockets of success with some clients using automation to drive the program, although this typically was not sustained in the business as my focus was on the technical side of things. Over the last 2–3 years, a new approach to data governance has emerged and is gaining traction focused on a crowd-sourced or collaborative approach to governance. In the companies that I've seen adopt this approach (including Airbnb and eBay), success had been sustained by allowing the community of analysts to govern themselves, with a layer of oversight for privacy and security to ensure regulatory compliance.

2. What makes an effective data governance program?

   I prefer the crowd-sourced approach driven by automation and collaboration capability that is gaining traction (see above). For this to succeed, it must also have a layer of oversight to ensure regulatory compliance.

3. Which industry sectors in your opinion have been most lax in implementing data governance and why?

   The public sector has relied on manual processes for data governance, resulting in large gaps in knowledge of systems and where data is being used. They appear to have adopted data governance recommendations some years ago but, due to financial constraints (budget and long buying cycles), they do not have the infrastructure in place to deal with data governance today. This is being slowly acknowledged but remediation programs are taking some time to be rolled out.

4. Which industry sectors have embraced data governance the most and what are their drivers for implementing data governance?

   Financial services companies have been investing in this space fairly heavily over the last 5+ years, seeking both technology and process solutions. Many of these organizations continue to struggle with manual process overhead, so I would expect them to follow the model adopted by Airbnb and eBay over the coming 24–36 months.

5. What is the role of data governance in achieving compliance?

   Data governance has a crucial role in compliance. It provides clarity for policies and proof of their application - which is essential for audit. It also provides confidence for the business to act without fear of inadvertently breaking rules or the law. Similarly, the same approaches, techniques and technology around data governance for compliance can be applied for profit as they shine a light on the business operations.

6. Any other interesting thoughts/advice that you would like to share with the audience of the book with respect to data governance?

   We will always need data governance but we will see much more of this done collaboratively or management from the bottom up rather than top-down, manually driven processes commonly used today.

7.9 Shannon Fuller, Director of Governance Advisory Services, Gray Matter Analytics

Shannon Fuller is the Director, Governance Advisory Services at Gray Matter Analytics, Charlotte, North Carolina, USA. He is a data governance thought leader and speaker. He assists healthcare clients in maturing their Information Governance and Analytics capabilities.

1. What are the biggest challenges that you face when implementing data governance?
   Cultural change is the biggest challenge. In many organizations, there is no culture of data ownership or active data management. When implementing a governance program, you must define your strategy based on your current culture.
2. How do you manage and resolve potential organizational conflicts from overlapping functions when dealing with enterprise data?
   Data ownership is ultimately about assigning accountability for making decisions about data. The domain owner must facilitate resolution, by gathering all requirements for the respective data from constituents across the organization.
3. How do you associate the data governance maturity levels with business benefits?
   The key is to demonstrate how data governance processes enable / support the strategic initiatives of the organization. In healthcare, most providers are moving to provide more on-demand care. That program cannot be successful without knowing who your providers are, what they can do and where they practice i.e., Mastering Provider.
4. What makes an effective data governance program?
   An effective program enables trust in information and risk reduction. Trusted information makes for information-based decision making and improves operational efficiency.
5. How has regulations (e.g., HIPAA) influenced data governance program in the healthcare sector and how has data governance helped meeting regulatory requirements?
   Regulations help to shape data governance policy and the implementation of those policies reduce the risk of violating those regulations. Regulations are a big stick that you have to wield lightly as a governance practitioner. You do not want data governance viewed as a compliance activity.
6. Any other interesting thoughts/advice that you would like to share with the audience of the book with respect to data governance?
   Your organization's culture will drive your governance strategy and your governance strategy will drive culture change.

7.10 Dr. Stan Rifkin, Director of R&D, Master Systems Inc.

Dr. Stan Rifkin is the founder and principal of Master Systems Inc. He worked at the Software Engineering Institute (SEI), the US Air Force, and the Systems Engineering Research Center, among others. He concentrates on deploying best practices to knowledge-creating enterprises of all sizes and objectives.

1. Why is data not treated as an enterprise asset?
   All enterprises have difficulty identifying their assets, and data, being intangible, and possibly fungible, is invisible, so it evades notice and attention.
2. Why do data governance programs fail?
   They fail for the reasons that any governance programs fail. The reasons are well documented and, unfortunately, experienced over and over again.
3. What makes an effective data governance program?
   The implementation of any/every technological change is a step-by-step planned initiative. It is not purely technical but rather takes into account organizational and human factors. I favor the mutual adaptation model: adapt the organization in light of the technology and adapt the technology in light of the particular organization.
4. In your opinion, how important is the chief data officer's role in the success of data governance?
   I am not sure. If the CDO has experience leading technological change, then it is important. If the CDO is siloed and inexperienced in making successful change, then you have to work around her.
5. How have regulations impacted data and influenced data governance?
   In the US federal sector, regulations are probably the governing driver. And in Europe, privacy regulations have shown light on the importance of tending to data.
6. Any other interesting thoughts/advice that you would like to share with the audience of the book with respect to data governance?
   I believe that one of the greatest challenges to successful data governance is inherent in its questions: who owns each datum, how is each one validated, and who is responsible the change of existence of each transformation and use? These are difficult questions and few organizations have the patience to thoughtfully answer them. So, data governance is stuck—and ignored.

7.11 Tony K. Epler, Chief Data Strategist, PricewaterhouseCoopers

Tony K. Epler is the Chief Data Strategist for PwC, responsible for leading and directing firm-wide data and information strategy and controls, designed specifically to leverage data for value, while identifying and remediating PwC issues and risks relevant to management of information as a strategic asset. He also serves as the

Head of Data Strategy and Policy for the newly created PwC Global Data Office, responsible for data management standardization in 157 countries.

1. In your opinion, why is there no standard definition around data governance?
   Lack of a standard definition for data governance can be attributed to two main causes. First is the fact that the responsibility, either designated or assumed, for data governance can reside in many different organizational functions within a business. In one company, it may reside with a specific data organization, while in other companies; data governance might be the responsibility of IT, or OGC or Risk and Compliance. It might even be assumed by such specific functions as Finance or HR or Sales and Marketing, depending on the nature of the business and the type of data.
   Clearly, each organizational function would have its own particular perspective abut data management, and thus, its definition of data governance. Second, an organizational definition of data governance can be highly dependent on the perception and use of data within the business.
   Heavily regulated industries such as health care or financial services are more concerned with things like compliance, risk reduction, standardization, data lineage, and the integrity of data flowing through the infrastructure. Less regulated industries, such as retail, must be able to react quickly to competition, so businesses are more concerned with agility, flexibility, and revenue generation than with risk and compliance. These radically different perspectives create different data strategies and processes and thus, different ideas around data governance.
2. How does data governance contribute to a data driven culture in your organization?
   PricewaterhouseCoopers (PwC) is a professional services organization. Data is the lifeblood of our firm. PwC does not manufacture automobiles or pharmaceuticals. We don't sell insurance or food items. Rather, we make and sell data. We take in data from clients, purchase third party data, and public open source data as our "raw material." Our practitioners then add unique PwC insight to the data to produce our "finished product" -- knowledge, advice, and wisdom. When serving our clients and our staff, data becomes shared knowledge for new and interesting business acumen. We expect our people to be able to access data anywhere, anytime, securely, and on any device. This enables us to develop a data-driven data-oriented culture which introduces data into the everyday conversation in the workforce and supports data sharing as central to firm success. Thus, data governance provides a framework for managing the people, process, policies, and culture around data.
3. Which business unit drives data governance in your organizations and why?
   PwC consists of three customer-facing business divisions—Assurance, Tax and Advisory. A fourth division, Internal Firm Services (IFS), contains the back-office functions such as HR, finance, legal, IT, etc.
   Previously, data governance was a decentralized assumed responsibility of the three business divisions. Many organizations across the firm had similar data

requirements but independently developed and deployed data-related solutions, technologies, and processes.

No organization or function had the responsibility for identifying, coordinating and deconflicting such initiatives. Four years ago, to bridge this gap and to provide the nexus to other functions in an effort to truly manage information as an asset, we concluded that mature information governance is a three-legged stool—the information has to be managed, it has to be protected, and value must be derived from the information, otherwise, it's of no use. Efficiently managing information could enable PwC to better meet legal and regulatory requirements while providing enhanced benefit and value to employees and customers.

For PwC to become a more data-driven organization, it was important to develop leading-edge capabilities in business intelligence, data mining, data analytics, and knowledge management in order to monetize data, provide more value to customers, and successfully grow the business. Protecting the security and privacy of both customer and internal information was critical to protecting the PwC brand. To accomplish all these goals, in a standardized, efficient, and cost-effective manner, and thus drive more value from data, we established the Office of the Chief Data Officer, assigned to the IFS division, with responsibility for centralized data governance. Fundamentally, the CDO is tasked with being the voice of data and generally representing data as a strategic business asset, concerned with where, when, how, and why the business uses data and transforms it into information that serves our customers. Over the past four years our CDO staff has grown from a skeleton staff of 4 to almost 60.

4. How do you measure the effectiveness of your data governance program?

   Like many organizations, we believe that programs, projects, and initiatives cannot be effectively managed unless they are measured. Consequently, we rigorously develop, assign, and track metrics and KPIs for all of our data governance actions and activities. We communicate these through dashboards built on high-end data visualization techniques and technologies. These dashboards are used internally by the CDO staff to manage the projects and initiatives. They are also accessed by the C-suite and executive staff to monitor and demonstrate ROI, organizational success, and compliance with overall firm strategy and business objectives.

5. How has GDPR influenced your data governance program?

   PwC is a global business with member firms in 157 countries. We have developed a comprehensive global data protection program and issued an extensive "toolkit" to help the member firms comply with GDPR provisions. The member firms then implement measures applicable to their particular countries and business goals.

   In the US, we have developed new policies, standards, and processes specific to GDPR compliance. Currently, we are analyzing other privacy initiatives, such as the new California law, to identify other emerging internal requirements.

6. Any other interesting thoughts/advice that you like to share with the audience of the book with respect to data governance?

Cross-industry studies show that on average, less than half of an organization's structured data is actively used in making decisions—and less than 1% of its unstructured data is analyzed or used at all. More than 70% of employees have access to data they should not, and 80% of analysts' time is spent simply discovering and preparing data.

Data governance is critical to ensuring that an organization can put forth clean, rightly protected data. Data governance establishes rules, processes, and controls to help an organization to know- where its data is? what it is? and what is the highest use of the data at any point in time? Data governance establishes that data issues are most often business process issues, rather than technical issues.

Appendix A
Bibliography

The more that you read, the more things you will know. The more that you learn, the more places you'll go. – Dr. Seuss, Renowned Children's Author, I can read with my eyes shut.

Smallwood R (2014) Information governance: concepts, strategies and best practices. Wiley, New Jersey.
Mahanti R (2019) Data quality: dimensions, measurement, strategy, management and governance. Quality Press, ASQ.
McDowall RD (2018) Data integrity and data governance: practical implementation in regulated laboratories. Royal Society of Chemistry.

R. Mahanti, *Data Governance and Compliance*,
https://doi.org/10.1007/978-981-33-6877-4

Appendix B
Glossary of Terms

Compliance

Compliance is about following laws and rules to prevent wrong things from happening as well as detecting and mitigating incorrect actions of the past.

Data

Data is a collection of facts related to real world entities, concepts and events, represented by text, numbers, images, sound or video, and formatted in a specific way to facilitate storage, processing, retrieval, analysis, sharing and usage.

Data governance

Data governance is the exercise and enforcement of policies, processes, guidelines, rules, standards, metrics, controls, decision rights, roles, responsibilities, and accountabilities to manage data as a strategic enterprise asset.

Data privacy

Data privacy relates to the ability of an organization or individual (to whom the data belongs) to determine what data can be shared with third parties and includes confidentiality requirements of the data.

Data security

Data security is the process of protecting data from any kind of action that can compromise the data such as unauthorized access, corruption, theft, and distribution.

Financial governance

The rigorous set of rules, guidance and policies issued by regulators to ensure that financial processes are well governed.

R. Mahanti, *Data Governance and Compliance*,
https://doi.org/10.1007/978-981-33-6877-4

Governance

The exercise of authority and control to ensure accountability and enhance transparency.

Governance, Risk and Compliance (GRC)

Strategy for organizational governance in way that risks are effectively managed while staying compliant with regulations. As the name suggests, GRC comprises of three constituents—governance, risk and compliance.

HR governance

The act of leading the HR function and managing related investments to optimize performance of the organization's human capital assets; fulfill fiduciary and financial responsibilities; mitigate enterprise HR risk; align the function's priorities with those of the business; and enable HR executive decision making (SABPP 2017; Mercer 2003).

IT governance

The processes that ensure the effective and efficient use of IT in enabling an organization to achieve its goals.

Operational governance

Operational governance is about collectively agreeing on how operating decisions will be made, executed and monitored and by whom.

Risk governance

The framework within which risk management functions in a corporation. Risk governance applies the principles of good governance to the identification, assessment, management and communication of risks (International Risk Governance Council).

Security governance

The security function dimension of the corporate governance framework and includes rules, policies, processes, roles and responsibilities, and decision rights to direct and control an organization's approach to protecting its physical and digital assets as well as protecting the organization's public reputation.

Index

A
Aaron Jones, 70
Accountability, 132
Adrian Cadbury, 18
Alan Westin, 123
American Assembly, The, 16
American Bar Association, 16
American Law Institute, 17
AnaCredit, 144
Andres Perez, 83, 160
Anne Marie Smith, 71
Anonymization, 123
Anthony Chalker, 72
Anti-money laundering, 114, 131, 136
Anti-Money Laundering and Counter-Terrorism Financing Act (AML/CTF Act), 136
    AML/CTF Act and KYC, 98
    Anti-Money Laundering and Counter-Terrorism Financing Act 2006 (AML/CTF Act), The, 98
    Anti-Money Laundering and Counter-Terrorism Financing Amendment Act 2017 (AML/CTF Amendment Act), The, 98
"Apply or Explain" approach, 39
Arthur Andersen, 22, 23
Asian currency crisis, 26
ASX Listing Rules, 33
Atedo Peterside, 41
Attributable, Legible, Contemporaneously recorded, Original or a true copy, and Accurate (ALCOA), 126
Australian Transaction Reports and Analysis Centre (AUSTRAC), 98
Australia's cash controls, 98
Authoritative source, 131

B
Bank of Ghana, The, 42
Barclays, 114
Barring Bank collapse, 30
Basel III, 105
Basel Committee on Banking Supervision's standard number 239 (BCBS 239), 100, 104, 105, 138
    principles for effective risk data aggregation and risk reporting, 100
BCBS 239 Principles, 100
    overarching governance and infrastructure principles, 101
        data architecture and IT infrastructure, 101
        governance, 101
    risk data aggregation capabilities principles, 102
        accuracy and integrity, 102
        adaptability, 102
        completeness, 102
        timeliness, 102
    risk reporting practices principles, 103
        accuracy, 103
        clarity and usefulness, 103
        comprehensiveness, 103
        distribution, 104
        frequency, 103
    supervisory review, tools, and cooperation principles, 104
        home/host cooperation, 104
        remedial actions and Supervisory measures, 104

R. Mahanti, *Data Governance and Compliance*,
https://doi.org/10.1007/978-981-33-6877-4

review, 104
Berle and Means, 14
Bhabha Committee, 29
Big data, 167
BNP Paribas S.A, 114
Border Force, The, 98
Borris Otto, 70
Bosch Report, The, 33
'Box-ticking' approach, 20
Business Roundtable, 16

C
Cadbury committee, 18, 19
Cadbury report, 18–20, 23, 36
Cape of Good Hope, 35
Capgemini, 71
Capital Requirements Directive, 105
implementing technical standards, 105
Carphone Warehouse, 114
CDO. *See* Chief Data Officer
Charles Roe, 72
Chief Data Officer, 162. *See also* CDO
Chief Privacy Officer (CPO), 132
China Regulatory Commission, 27
Christopher Butler, 69, 161
Code of Best Practice, 19
Cohen (2006), 70
Combined Code of Corporate Governance, 22
Combined Code, The, 20
Commercial Code, 32
2002 Commercial Code reform, 32
Committee of Sponsoring Organizations, 18
Common Reporting (COREP), 105, 106
2013 Companies Act in India, 30
Companies Act, The, 29, 31, 32
1968 Companies Act, The, 40
Companies Ordinance, The, 29
Company Law Forum, 23
Company legislation, 29
Compliance, 5, 67, 80, 85, 109–111, 119, 121, 122, 126, 131–134, 136, 142, 144–147, 149, 150, 157
'tick box' approach, 150
change management, 134
compliance and performance, 5, 6, 84, 85, 150
compliance challenges, 120
compliance data, 143
cost of compliance versus cost of non-compliance, 118
cost of non-compliance - examples, 116, 117
data characteristics and compliance, 122, 146
data accessibility, 124
data accuracy, 125
data completeness, 125
data consistency, 125
data currency and timeliness, 126
data granularity, 127
data integrity, 126
data privacy, 123
data protection and privacy laws and data characteristics, 128
data retention, 127
data security, 123
data usage, 127
traceability, 126, 127
data granularity, 144
data management, for, 142
data requirements around compliance, 111
data retention and compliance, 127
role of data governance in compliance, 112
stakeholders and compliance, 134
Compliance data, 145–147
single view of, 145
Compliance data store, 145–147
compliance data mart, 137, 145
compliance data store design, 146
data management activities, 147
Compliance, Governance and Oversight Council, 118
"Comply or Explain" approach, 39, 40
Confederation of Indian Industry, 29
Controls, 74, 79
Corporate accounting, 140
Corporate data warehouse, 145
Corporate Director's Guidebook, 16
Corporate governance, 11, 13, 15, 16, 21, 23, 26, 33, 35, 38, 39, 43, 51, 53–55, 57, 59, 76, 77, 81
board, 3
corporate governance - definition, 3, 11
corporate governance framework, 53
corporate governance practices, 10
corporate governance subdisciplines, 53
facts and corporate governance, 5
governing body, 2
no "one-size-fits-all" approach, 53
OECD Principles of Corporate Governance, 11
subdisciplines of corporate governance, 4

subset disciplines of corporate governance, 52
Corporate governance and data governance, 76
Corporate governance and HR governance, 56
Corporate governance code, 17, 28, 32
Corporate Laws Amendment Act, 36
Corporate scandals, 17
Corporate strategy, 53, 81, 156
Corporate vision and mission, 81
Corporation, 14
Corporations Act, 33, 34
COSO framework, 77, 78
control activities, 78
control environment, 78
information and communications, 78
monitoring, 79
risk assessment, 78
COSO Report, 18
Cost of compliance, 114–116
Cost of non-compliance, 118
cost of non-compliance - examples, 114. *See also* Non-compliance costs
Cost of non-compliance - examples, 114
1998 Data Protection Act - cost of non-compliance, 114, 115
1998 Data Protection Act non-compliance -example, 114
AML non-compliance cost- examples, 117
HIPAA fines, 116
HIPAA Privacy and Security Rules violation, 116
Counterparty risk assessment, 140
CRD IV, 105
Critical data, 125, 131
Critical data element quality, 111
Cultural change, 176
Culture, 173
Current Good Manufacturing Practices (cGMP), 126
CVS Health, 65, 83, 172

D

DAMA International, 70
Dan Power, ix–xi, 76
Data, ix, 5, 63, 65, 134, 145, 160, 164, 166, 167, 173, 178
life cycle, 74
multiple versions of truth, 145
single view, 138
Data archiving, 127
Data breach, 115
Data characteristics and compliance - interrelationships, 128
Data characteristics - interrelationships, 128
Data classification, 63, 64, 127, 144
by data sensitivity and protection, 64
Data consumers, 74
Data/data quality strategy, 81
Data-driven, 173, 179
Data elements, 143, 145
Data governance, ix, 63, 65–67, 74, 76, 77, 81, 83, 84, 110–112, 120, 121, 131–133, 136–139, 141–146, 162–164, 176, 178, 180, 183
business case for, 163
crowd-sourced approach, 175
culture, 165
culture change, 173
data governance aspects, x, 72
data governance benefits, 157, 170
data governance - definition, 67
data governance definition - lack of standardization, 68
data governance definitions - literature review, 70
data governance drivers, 150, 157
data governance framework, 165
data governance maturity, 163
data governance operating models, 164
data governance programs, 72
components, 72
defensive data governance, x
metrics, 133, 168, 174
misconceptions, 69
offensive data governance, x
organization structure, 74
people aspect and organization structure, 75
technology, 76
Data governance and compliance, 150, 175
accountability, 122
data governance, AML/CTF and KYC, 136
data governance and BCBS 239, 139
data governance and Dodd-Frank Act, 141
data governance and FATCA, 140
data governance and GDPR, 138
data governance and PCI DSS, 139
data governance and regulatory reporting, 137
data governance and SoX, 140

data governance, FINREP and COREP, 137
five Ws and one H, the, 129
HIPAA and data governance, 65
ownership, 122
working committees, 121
Data governance and data management, x
Data governance and data management functions, 69
Data governance and GRC, 81
Data Governance Australia, 72
Data Governance Institute, The, 70
Data governance policy, 133
Data granularity, 127, 137, 144
Data integrity, 126
Data landscape, 145
Data lineage. *See* Traceability, 130
Data management disciplines and data initiatives, 157
Data ownership, 145, 176
Data phenomena, 84, 151
Data policies, 74, 111, 133
Data privacy, 124
Data privacy and security, 131
Data privacy versus data security, 124
Data producers, 74
Data profiling and assessment, 143
Data protection, 120
Data protection laws, 120
Data Protection Officer (DPO), 132
Data protection versus data security, 123
Data quality, 132
assessment, 143–145
controls, 132
metrics, 147
standards, 145
Data remediation, 143, 144
Data requirements and compliance, 121
data discovery, 119, 131
data in context of compliance, 134
single view of data, 136
Data retention, 121, 127
Data risks, 81
Data security, 124
Data security governance, 59
Data standardization, 147
ISO 20022, 147
Data standards, 125, 131, 147
Data steward, 160
Data stewardship, 145
Data strategy, 156
Data swamp, 167
Data usage, 127
Data usage policies, 127
Data value realization, 161
Data vision and mission, 81
David Loshin, 70
David Walker, 22
De Haes and Van Grembergen, 60
De-Identified Health Information, 90
Derek Higgs, 21
DG. *See* Data governance, 166
Digital age, 5
Dodd-Frank Act, 25, 26, 141
Dutch East India Company, 35

E

Economic globalization, 35
Elait Australia, 174
Electronic processing, 65
Electronic Protected Health Information (e-PHI), 91
Elizabeth Pierce, 72
Encryption, 124
English Companies Act, 42
English law, 35
Enron, 22, 23
Enterprise data repository, 66
Entity, 1
Equifax, 115
Ernst & Young, 23
European Banking Authority (EBA), 105, 125
European Union, 23, 120, 138
Evan Levy, 70
Evolution of corporate governance, xiv, 12, 15
Cadbury committee, 18
corporate governance dispute, 14
in Africa, 35
in Asia, 26
in Australia, 33, 34
in China, 26, 27
in Europe, 23
in Ghana, 42
in Hong Kong, 28, 29
in India, 29, 30
in Japan, 31–33
in Kenya, 38–40
in Nigeria, 40, 41
in Singapore, 30, 31
in South Africa, 35–38
in UK, 18, 21, 22
in USA, 15–18, 23
key milestones in corporate governance evolution, 156

origin of corporate governance, 14
World War II, 31
Evolution of data governance, 65–67, 69
Evolution of governance, 5
ancient scriptures and governance, 12, 13
Arthashastra, 13
Bhishma, 13
Chanakya, 12, 13
Dharma, 13
Qur'an, 12
Raj Dharma, 13
Shanti Parva, 13
Chandragupta Maurya, 12
Confucian ideal, 12
Darius I, 12
King, The, 12
Mauryan empire, 12
Tang Taizong, 12
Exchanges Act, 30
Experian, 72

F
Facebook, 115
Fair Labor Standards Act, 114
Family capitalism, 33
FATCA data requirements, 99
Federal Food, Drug, and Cosmetic Act (FDA), 113
Financial CHOICE Act, 26
Financial governance, 55
financial governance and financial crisis, 55
Financial Reporting (FINREP), 105, 106, 140
Financial reporting regulations, 156
Financial Services Authority, 106
First Sans Francisco Partners, 71
Foreign Account Tax Compliance Act (FATCA), 99, 140
Foreign and Corrupt Practices Act, 16
Foreign Financial Institution (FFI), 99
Forrester Research, 120
Franklin D. Roosevelt, 14
FRC guidance, 21

G
Gartner, 60
GDPR, 97, 120. *See also* General Data Protection Regulation
controller, 92, 94
cost of GDPR non-compliance - examples, 115
Data Protection Directive (95/46/EC) versus GDPR, 92
data retention policy and GDPR, 128
GDPR act, 92
GDPR concerns, 120
GDPR principles, 93
accountability, 94
accuracy, 94
data minimization, 94
integrity and confidentiality (security), 94
purpose limitation, 94
storage limitation, 94
GDPR – Two Tiers of Fines, 92
personal data, 94
personal data collection and processing, 94
personal data processing - lawful bases, 92
consent, 93
contract, 93
legal obligation, 93
legitimate interests, 93
public task, 93
vital interests, 93
personal data storage, 94
privacy by default, 95
privacy by design, 94
processor, 92
PwC and GDPR, 179
Rights of Individuals under GDPR, 95
automated individual decision-making, 97
profiling, 97
right of access, The, 96
right to be forgotten, The, 96
right to be informed, The, 95
right to data portability, The, 97
right to erasure, The, 96
right to object, The, 97
right to rectification, The, 96
right to restrict processing, The, 96
rights in relation to automated decision making and profiling, 97
General Data Protection Regulation. *See* GDPR
George Firican, 65, 163
Ghanaian Companies Code, 42
Ghana SEC, The, 42
Ghana Stock Exchange, 42
Global financial crisis, 55, 57, 67, 105
Governance, 1, 2, 9, 10, 12, 79, 161, 171
governance - definition, 10

governance strategy and culture, 176
Governance and management, 1, 2
Governance, risk, and compliance. *See* GRC, 79
Governance versus management, 1, 2
Robert Tricker, 2
Gray Matter Analytics, 150, 176
GRC, 80. *See also* Governance, risk, and compliance
GRC benefits, 81
GRC components, 79
Great depression, The, 14
Greenbury committee, 19
Greenbury Report, 19, 20
Gwen Thomas, 70

H

Hampel committee, 20, 21
Hampel Report, 20
Harold Wilson, 15
Health Care Clearinghouses, 90
Health Care Providers, 90
Health Insurance Portability and Accountability Act (HIPAA), 65, 90, 128, 173, 174
business associate, 90
Health Plans, 90
Higgs report, 21
HIH Insurance, 34
HIH Royal Commission, 34
Hilton, 115
HIPAA
HIPAA Breach Notification Rule, 91
HIPAA Privacy Rule, 91
HIPAA Security Rule, 91
HIPAA. *See* Health Insurance Portability and Accountability Act (HIPAA)
HR governance, 56, 57. *See also* Human resource governance
Sarbanes–Oxley - HR implications, 56
HSBC, 114
HSBC UK, 69
HSBC, UK Bank plc, 161
Hugo Young, 15
Human resource governance. *See* HR governance
Human resources, 56

I

IBM Data Governance Council, 71
Identity access management tools, 124
Individually identifiable health information, 91
Informatica, 72
Information Technology governance. *See* IT governance
Institutional capitalism, 33
International Association of Privacy Professionals, 123
International Emergency Economic Powers Act, 114
Internet, 66
IRM Consulting Ltd. Co., 83, 160
IT governance, 59, 60, 63
board and IT governance, 63
Chief Executive Officer, 63
Chief Information Officer, 63
IT governance - definition, 60
IT governance frameworks, 60
COBIT framework, 79
Control Objectives for Information and Related Technology (CobiT), 60
Factor Analysis of Information Risk (FAIR), 62
ISO/IEC 38500, 62
IT Infrastructure Library (ITIL), 62
Value from IT Investments (Val IT), 62
objectives, 60
IT Governance Institute, 60
IT governance versus data governance, x, 78

J

Jan van Riebeeck, 35
Japanese Commercial Code, 31
Japanese corporate governance model, 31
Japanese Sarbanes Oxley Act, 32
Japan's Corporate Governance Code, 32, 33
Japan's Stewardship Code, 32
Jill Dyché, 69, 70, 164
Jill Dyché, LLC, 69
John A. Zachman, 3, 171
John Ladley, 70
John R. Talburt, 83, 166
J.P. Morgan Chase Bank, The, 114

K

Kansayaku, 31
Khatri & Brown, 2010, 71
King Committee, The, 35, 36
King report, The, 36
entities, 36
King I report, The, 36

key principles, 36
King II report, The, 36
key principles, 36
King III report, The, 37
new principles, 37
King IV report, The, 37
key principles, 37
Know Your Customer (KYC), 97, 99, 125, 131, 136
KPI, 174
Kumar Mangalam Birla Committee, 29
recommendations, 29

L

Laura Sebastian-Coleman, 65, 83, 172
Laws, 111, 119, 120, 127
Legal entity identifiers, 147
Legal requirements, 119
Legislation, 119–121
Limited Liability Act of 1855, 14
Logging, 147
London Stock Exchange, 19, 20
Lord Turner, 22

M

Management, 1–3
Managerial capitalism, 33
Margaret Rouse, 69, 72
Maria Villar, 70
Masking, 124
Master data, 64
Master Systems Inc., 177
MDM Institute, The, 70
Melvin Eisenberg, 16
Mervyn E. King, 36
Metadata, 64, 131, 169
Metaphysical stage, 84
Metrics, 74
Microsoft Excel, 168
Modern Corporation and Private Property, The, 14
Monetary Authority of Singapore, 30
Money laundering, 98
Morgan Templar, 71

N

Napoléon Bonaparte, 63
National Institute of Standards and Technology (NIST), 70
Newman and Logan (2006), 70
New York Times, 16
Nigel Hawthorn, 120
Nigel Turnbull, 21
Noetic Partners Inc, 83
Non-compliance costs. *See* Cost of non-compliance
Non-compliance in regulated industries-impacts, 112

O

Old Code, The, 28
OMES Data Governance Program Office, 71
Operational governance, 55
Oracle, 71
Organizational conflict, 173
Organization for Economic Co-operation and Development (OECD), 34
Ownership, 132

P

Pan El crisis, 30
Payment Card Industry Data Security Standard (PCI DSS), 106, 107, 139
Payment data, 106
PCI Compliance, 107
PCI Security Standards Council, 106
PDC layer, 167
Penn Central, 15
Performance, 85
Performance and conformance, 2
Personal health information, 90, 128
Personally Identifiable Information (PII), 124, 142
Phil Watt, 112, 174
Policies, 55, 127
Ponemon Institute and Globalscape, 117
Positive stage, 84
PricewaterhouseCoopers (PwC), 67, 68, 71, 118, 177, 178
Privacy, 123
Privacy information, 95
Protected Health Information (PHI), 90, 91
Purging, 127

R

Ralph Nader, Mark Green, and Joel Seligman, 15
Ramsay Report, The, 33
Reference data, 64
Regulations, ix, 67, 76, 79–81, 89, 110, 111, 118–120, 122, 126, 127, 131–134,

144, 146, 147, 149, 156, 157, 162, 176, 177
regulatory drivers, 113
Regulations and data, 118
data and regulations – challenges, 119
Regulations and the role of data governance, 134
Regulatory compliance and data management, 141, 147
Regulatory documents, 141
Regulatory report, 130
Regulatory reporting, 125, 128, 136, 145
Regulatory reporting and data characteristics, 128
Regulatory requirement, 113, 130
Regulatory security requirements, 144
Reporting entities, 98
Revised Code of Corporate Governance, 41
Richard Eells, 15
Richard Greenbury, 19
Right to be forgotten, 121
Risk culture, 58
Risk data aggregation, 100, 139
Risk data aggregation capabilities, 100
Risk governance, 57, 58
Board's role, 58
risk governance framework, 58
Risk management, 57, 58, 102
Risk reporting practices, 102
Risk reports, 103
Risks, 57, 58
Risks and industry sector, 57
Risk types, 100
Robert Seiner, 70
Robert Smith, 22
Robert Stroud, 120
Robert Tricker, 15
Roman-Dutch law, 35
Ronald Hampel, 20
Root cause analysis, 143
Royal Bank of Scotland, The, 114
Rules and standards, 74

S

Sarbanes–Oxley, 24, 25
Sarbanes–Oxley Act (SOX), 24, 25, 77, 79, 140
SEC 2003 code, 41
Securities Act of 1933, 14, 15
Securities and Exchange Commission, 140
Securities Exchange Act of 1934, 14, 15
Securities Exchange Commission, 15
Securities Industry Law, 42
Security controls, 145
Security governance, 58, 59
security governance framework, 59
Segregation of duties, 144
Self-Assessment Questionnaire, 107
Sensitive data, 145
Shannon Fuller, 133, 150, 176
Singapore Exchange, 30
Singapore Institute of Directors, 30
Singapore Stock Exchange, 30
Skyhigh Networks, 120
Smith Report, 22
Social Security Number, 143
SOE governance, 27
Solvency II, 77
SoX compliance, 140
Spiceworks, 120
Stakeholders, 149
Standardization, 135
Standards Australia, 34
Stan Rifkin, 113, 177
State-Owned Enterprises (SOEs), 26, 35
Statutory auditor, 31
Steve Sarsfield, 70
Stewardship Code, The, 32, 33
Stewart Kyd, 14
Stock Exchange of Hong Kong, 28
Strategy, 163
Sunil Soares, 70

T

TalkTalk, 115
Theological stage, 83
Timeliness, 126
Timestamps, 126
Toba Beta, 51
Tokyo Stock Exchange, 32
Tony Epler. *See* Tony K. Epler
Tony Fisher, 70
Tony K. Epler. *See also* Tony Epler
Tony King, 144
Traceability, 126
Trading with the Enemy Act, 114
Transactional data, 64
Treadway Commission, 18
Treadway report, 18
Turnbull Report, 21
Turner Review, 22
Tyson Report, 22

U
UBC DAE, 65
UK Financial Reporting Council, 22
UK Listing Rules, 20
UK Stewardship Code, 32
Union Companies Act, The, 35
University of Arkansas, 83, 166
User access list, 144
    assessment of, 145

W
Walker Review, 23
Watergate scandal, 16
Wikipedia, 71
1989 World Bank report, 35
WorldCom, 24

X
XBRL reporting, 105

Z
Zachman Framework™, The, 171
Zachman International, 171
Zaibasu, 31
Zeljko Panian, 71

GPSR Compliance
The European Union's (EU) General Product Safety Regulation (GPSR) is a set of rules that requires consumer products to be safe and our obligations to ensure this.

If you have any concerns about our products, you can contact us on

ProductSafety@springernature.com

In case Publisher is established outside the EU, the EU authorized representative is:

Springer Nature Customer Service Center GmbH
Europaplatz 3
69115 Heidelberg, Germany

www.ingramcontent.com/pod-product-compliance
Lightning Source LLC
Chambersburg PA
CBHW070112150825
31149CB00003BA/27
* 9 7 8 9 8 1 3 3 6 8 7 6 7 *